Common Core Math
For Parents

FOR

DUMMIES

A Wiley Brand

by Christopher Danielson, PhD

FOR

DUMMIES

A Wiley Brand

Common Core Math For Parents For Dummies®

Published by
John Wiley & Sons, Inc.
111 River Street,
Hoboken, NJ 07030-5774,
www.wiley.com

Copyright © 2015 by John Wiley & Sons, Inc., Hoboken, New Jersey

Published simultaneously in Canada

No part of this publication may be reproduced, stored in a retrieval system or transmitted in any form or by any means, electronic, mechanical, photocopying, recording, scanning or otherwise, except as permitted under Sections 107 or 108 of the 1976 United States Copyright Act, without the prior written permission of the Publisher. Requests to the Publisher for permission should be addressed to the Permissions Department, John Wiley & Sons, Inc., 111 River Street, Hoboken, NJ 07030, (201) 748-6011, fax (201) 748-6008, or online at http://www.wiley.com/go/permissions.

Trademarks: Wiley, For Dummies, the Dummies Man logo, Dummies.com, Making Everything Easier, and related trade dress are trademarks or registered trademarks of John Wiley & Sons, Inc., and may not be used without written permission. All other trademarks are the property of their respective owners. John Wiley & Sons, Inc., is not associated with any product or vendor mentioned in this book.

LIMIT OF LIABILITY/DISCLAIMER OF WARRANTY: WHILE THE PUBLISHER AND AUTHOR HAVE USED THEIR BEST EFFORTS IN PREPARING THIS BOOK, THEY MAKE NO REPRESENTATIONS OR WARRANTIES WITH RESPECT TO THE ACCURACY OR COMPLETENESS OF THE CONTENTS OF THIS BOOK AND SPECIFICALLY DISCLAIM ANY IMPLIED WARRANTIES OF MERCHANTABILITY OR FITNESS FOR A PARTICULAR PURPOSE. NO WARRANTY MAY BE CREATED OR EXTENDED BY SALES REPRESENTATIVES OR WRITTEN SALES MATERIALS. THE ADVISE AND STRATEGIES CONTAINED HEREIN MAY NOT BE SUITABLE FOR YOUR SITUATION. YOU SHOULD CONSULT WITH A PROFESSIONAL WHERE APPROPRIATE. NEITHER THE PUBLISHER NOR THE AUTHOR SHALL BE LIABLE FOR DAMAGES ARISING HEREFROM.

For general information on our other products and services, please contact our Customer Care Department within the U.S. at 877-762-2974, outside the U.S. at 317-572-3993, or fax 317-572-4002. For technical support, please visit www.wiley.com/techsupport.

Wiley publishes in a variety of print and electronic formats and by print-on-demand. Some material included with standard print versions of this book may not be included in e-books or in print-on-demand. If this book refers to media such as a CD or DVD that is not included in the version you purchased, you may download this material at http://booksupport.wiley.com. For more information about Wiley products, visit www.wiley.com.

Library of Congress Control Number: 2014954677

ISBN 978-1-119-01393-8 (pbk); ISBN 978-1-119-02112-4 (ebk);
ISBN 978-1-119-02111-7 (ebk);

Manufactured in the United States of America

10 9 8 7 6 5 4 3 2 1

Contents at a Glance

Introduction *1*

Part I: Getting Started with
Common Core Math Standards *5*

Chapter 1: The Lowdown on Common Core Math, Just
the Basics 7
Chapter 2: Looking at Math Teaching Then and Now 17
Chapter 3: Exploring the Standards for Mathematical Practice 25
Chapter 4: Understanding Homework Assignments 37

Part II: Focusing on Elementary Math:
Kindergarten through Fifth Grade *45*

Chapter 5: Beginning with Kindergarten Math 47
Chapter 6: Solving Problems in First-Grade Math 57
Chapter 7: Prioritizing Place Value in Second Grade 69
Chapter 8: Finding Fractions in Third-Grade Math 79
Chapter 9: Mastering Multiplication in Fourth-Grade Math 93
Chapter 10: Anticipating Algebra in Fifth-Grade Math 105

Part III: Moving Up to Middle and High School
Math: Sixth through Twelfth Grade *119*

Chapter 11: Relating to Ratios in Sixth-Grade Math 121
Chapter 12: Pursuing Proportions in Seventh-Grade Math 137
Chapter 13: Arriving at Algebra in Eighth-Grade Math 151
Chapter 14: Looking at Advanced Math: High School
and Beyond 167

Part IV: The Part of Tens *185*

Chapter 15: Ten Awesome Resources for Parents 187
Chapter 16: Ten (or So) Proven Ways to Support
Math at Home 191

Index *195*

Table of Contents

Introduction.. 1

About This Book 1
Foolish Assumptions 2
Icons Used in This Book.. 3
Beyond the Book... 4
Where to Go from Here 4

*Part I: Getting Started with
Common Core Math Standards*............................... 5

**Chapter 1: The Lowdown on Common Core Math,
Just the Basics** 7

Understanding What Common Core Math Is........................ 8
Examining the Standards for Mathematical Practice............ 9
    Ask questions... 10
    Play 10
    Argue 10
    Connect 11
Looking at the Standards in the Different Grades 11
    Kindergarten.. 12
    First grade... 12
    Second grade... 13
    Third grade.. 13
    Fourth grade.. 13
    Fifth grade... 14
    Sixth grade.. 14
    Seventh grade.. 15
    Eighth grade 15
    High school.. 15
Helping Your Child with Homework..................................... 16

**Chapter 2: Looking at Math Teaching
Then and Now** 17

Setting Goals for the 1900s 17
Competing Globally with Advanced Math and Science 18
Returning to Basics... 19

Teaching Math in the Info Age 20
Holding Teachers Accountable 21
Agreeing on the Common Core 23

Chapter 3: Exploring the Standards for Mathematical Practice 25

Focusing on Asking Questions 26
    Asking "Why?" and "How do you know?" 26
    Asking "What if?" 27
    Asking "Is it good enough?" 28
    Asking "Does this make sense?" 29
    Asking "What's going on here?" 30
Playing with Math 32
    Experimenting with symbols 32
    Building models 33
Arguing Is a Ton of Fun 34
Connecting Ideas 35

Chapter 4: Understanding Homework Assignments 37

Getting the Inside Scoop: Teachers' Views on
    Homework 37
Examining the Homework Assignments on
    Social Media 39
        Number lines 39
        Dots 41
Becoming Unstuck: What to Do 42
Helping Your Child without Doing the Work Yourself 43

Part II: Focusing on Elementary Math: Kindergarten through Fifth Grade 45

Chapter 5: Beginning with Kindergarten Math 47

Counting by Ones and Tens 47
    Representing numbers: Counting and
        cardinality 48
    Comparing numbers: More, less, and equal 49
Focusing on Operations and Algebraic Thinking 50
    Solving problems by counting 50
    Putting numbers together and taking
        them apart 51
Noticing Ten: Place Value 52

Comparing and Classifying: Foundations
of Measurement.. 52
Getting Started with Geometry 53
Describing shapes 54
Playing with shapes.. 55

Chapter 6: Solving Problems in First-Grade Math...............................57

Digging in to Addition and Subtraction................ 57
Saying bye-bye to keywords.............................. 58
Decomposing numbers 61
Focusing on the Decimal Number System:
Place Value Begins 62
Grasping the significance of 10......................... 62
Representing tens... 63
Working with All Sorts of Measurements 65
Measuring length 65
Working with time and money.......................... 66
Measuring the concrete: Data........................... 67
Delving into Basic Geometry................................. 67

Chapter 7: Prioritizing Place Value in Second Grade 69

Adding, Subtracting, and Looking for Groups...... 70
Thinking about addition and subtraction.......... 70
Seeing why units matter 72
Focusing on Place Value 74
Going Deeper with Measurement 76
Measuring length 76
Working with time and money.......................... 77
Identifying and Building Shapes............................ 78

Chapter 8: Finding Fractions in Third-Grade Math...............................79

Studying Multiplication and Division 79
Defining multiplication and division 80
Tackling properties 82
Mastering Addition and Subtraction 85
Exploring Fractions... 87
Partitioning Cutting up things........................... 87
Putting fractions on number lines..................... 89
Estimating and Measuring Precisely: Having

It Both Ways... 90
    More than just rounding... 90
    Counting squares to find area....................................... 91
    Representing data with graphs...................................... 92
Categorizing Shapes 92

Chapter 9: Mastering Multiplication in Fourth-Grade Math..........................93

Focusing on Multiplication: Factors and Multiples 93
Calculating with Place Value 94
    Decomposing numbers 95
    Striving for fluency: The case of addition
        and subtraction... 95
Forming Fractions... 96
    Using unit fractions 96
    Going beyond reducing: Equivalent fractions............ 97
    Comparing fractions.. 99
Eyeing Units and Angles.. 101
    Mastering units of measure...................................... 101
    To every angle turn, turn, turn 102
Addressing Lines and Angles 103

Chapter 10: Anticipating Algebra in Fifth-Grade Math 105

Expressing Relationships: Early Algebra Ideas.................. 106
    Generalizing: The beginnings of algebra.................... 107
    Please Excuse My Dear Aunt Sally:
        Order of operations.. 108
Extending Place Value and Algorithms 111
    Placing value: Left and right of the
        decimal point... 111
    Standard algorithms: Doing things the
        old-fashioned way.. 112
Operating on Fractions 114
    Thinking about fractions 114
    Adding, subtracting, and multiplying
        fractions 115
Measuring Volume and Graphing Data 117
    Calculating volume: The number of cubes
        that fit inside 117
    Using a quick and dirty graph: The line plot............ 118
Concentrating on Properties of Shapes 118

Part III: Moving Up to Middle and High School Math: Sixth through Twelfth Grade 119

Chapter 11: Relating to Ratios in Sixth-Grade Math . . 121

Understanding Ratios 122
  Telling fractions from ratios.................................... 122
  Solving ratio and rate problems 123
Examining Multiplication 125
  Differentiating between GCFs and LCMs 125
  Dividing fractions can be fun 126
Extending from Arithmetic to Algebra 130
  Using variables.. 130
  Figuring out equivalence 131
Measuring Area and Volume 131
  Focusing on area.. 132
  Getting 3-D with volume.. 134
Measuring Datasets 135
  Eyeing measures of center 135
  Focusing on spread 136

Chapter 12: Pursuing Proportions in Seventh-Grade Math 137

Examining Ratios and Proportional Relationships 138
  Keeping track of key terms...................................... 138
  Seeing how proportions are key 139
  Solving proportions.. 141
Working with Negative Numbers 142
  Multiplying and dividing with properties................. 143
Describing Things with Algebra...................................... 145
Delving Deeper into Geometry....................................... 146
  Measuring circles: A piece of pi 147
  Tackling triangles 147
Studying Data and Chance... 149
  Considering: basic probability................................ 149
  Making sense of data: Statistics 150

Chapter 13: Arriving at Algebra in Eighth-Grade Math........................ 151

Tackling Irrational Numbers... 151
Using Exponents Equations 152
  Looking at linear relationships 153
  I got the power! Using exponents 156
  Solving two equations at once................................. 157

Delving into Functions..159
Defining functions.......................................159
Figuring out a function's function............................161
Doing the Ancient Greeks Proud..162
Deciding what's the same............................162
Measuring the hypotenuse.......................................163
Finding tricky volumes.............................165
Addressing Bivariate Data 166

**Chapter 14: Looking at Advanced Math:
High School and Beyond.....................167**

Knowing What College Ready Means 167
Examining some truths about college
math placement 168
Looking at integrated math 170
Advancing an Algebraic Agenda 172
Studying function families 173
Solving systems................................174
Using algebraic structure 176
Getting Formal with Geometry...177
Proving stuff 177
Using similarity and trigonometry.........................178
Describing shapes with algebra.........................179
Understanding the World, Statistically Speaking 180
Quantifying data in one or two variables 181
Understanding the relationship between data
and probability...182
Using probability to make decisions.........................183

Part IV: The Part of Tens............................... 185

Chapter 15: Ten Awesome Resources for Parents . . .187

Talking Math with Your Kids...............................187
Moebius Noodles 188
Estimation 180...188
Visual Patterns 188
Math Educators Stack Exchange............................189
Math Forum 189
YouCubed 189
Your Public Library 189
Math Munch..190
Common Core State Standards Initiative...........................190

Chapter 16: Ten (or So) Proven Ways to Support Math at Home **191**

Talking about Math Together.............................. 191
Perusing Your Child's Notebook......................... 192
Explaining a Solved Problem.............................. 192
Playing a Board Game 193
Having Number Talks 193
Grocery Shopping Together 193
Arguing about Words 194
Building Things 194
Asking "What If?".. 194

Index.. *195*

Introduction

*M*ath shouldn't be scary. This idea is at the heart of this book. The big reason that many people find math scary (and I know you're out there — you confess it to me when we meet for the first time and I tell you that I teach math) is that math has always felt like one big club with a bunch of rules that make no sense, but that absolutely must be followed.

This book presents a very different vision of math — one that should be empowering rather than frightening. A major goal of the Common Core State Standards is raising the mathematics achievement of large populations of students to whom quality mathematics instruction has previously been denied. Part of this effort involves bringing children's ways of thinking into the classroom and developing these ways of thinking into powerful, useful, and efficient strategies. I wrote *Common Core Math For Parents For Dummies* to help parents like you understand this development.

I have been writing about students' math learning at all levels for many years now, with an audience of both parents and teachers. With this book, I want to bring together many of these ideas into a coherent and comprehensive guide to classroom math learning. *Common Core Math For Parents For Dummies* is that book. So go ahead and join in — the mathematical welcome mat has been laid out for you.

About This Book

This book is your guide to math class in the 21st century. Education policy can be highly political and contentious, so this book cuts through it all and tells you what you need to know about what and how your child is likely to be learning math in the era of Common Core.

In place of inflated claims about the perfect world that will supposedly result from adopting these standards, you can find honest information about the goals and intent of these standards. Instead of scary tales of data mining and big government, you can find reasonable, measured, and careful descriptions of what the standards actually are.

If you're a parent or guardian, you can find suggestions for helping your children learn the math appropriate to their grade level. This information may take the form of written tips for working example problems or video explanations of important ideas. Furthermore, this book shows you how ideas you learned in school are likely to appear in your children's math class. Despite what you may have heard, the standards don't have any New Math in them. Children are being asked to think, and this thinking can look unfamiliar to their parents on the surface. But underneath, many of these ways of thinking are old and familiar. Many times people who identify themselves as "not a math person" will say something such as, "That's how I always thought about it, but I didn't know it was okay to do it that way!" This book can help you connect your child's ways of thinking with your own.

If you're a teacher, you can find a most welcome big picture. You can see connections between the math that you teach at your grade level (which you probably know quite well) and the math that is taught at adjacent and distant grade levels (which you probably haven't had time to study).

This book is organized as a reference that you can spend as little or as much time with as you want. You can read the grade level that matters to you without worrying about what came before and what comes after. All in all, I wrote this book with a busy person in mind. I have organized things so you can find what you need and move on.

Foolish Assumptions

As I wrote this book, I made some assumptions about you. I'm sure I didn't get them all right, but at least one of these categories describes you:

✔ **You have a K–12th grade student in your life.** You may be a parent, guardian, grandparent, neighbor, or tutor to a child you care about very much, and you want to help him (or her) be successful in math in school.

✔ **You don't really know what Common Core Math means.** You have probably heard of the Common Core State Standards, but you probably haven't read them.

✔ **You have seen something unfamiliar in your child's homework.** Seeing something that you thought you knew well (for example, multiplication) but realizing that you have no idea what the questions are asking for can be frustrating.

✔ **You are a teacher looking to know the standards better.** Understanding the standards beyond the grade level you teach is extremely helpful for day-to-day classroom teaching. (How did they learn this last year? How will this get used in high school?) It's also helpful in supporting parents when they have questions. Either way, you need information quickly.

I understand that your life is busy. I wrote this book in a way that makes the phrase Common Core Math concrete. The goal is to bring you up to speed quickly on what Common Core means for your child's math class.

Icons Used in This Book

Throughout this book, I include icons in the margins. You can use these icons to navigate this book.

A tip is intended to make your life easier. A tip can give you suggestions of what to look for in the standards or in your child's work.

This icon helps you find summaries of the most important ideas in a section. This icon points to something that you won't want to forget.

This icon lets you know when you can do something with your child in order to understand the content and help your child. You can do some of them on your own as you read; others suggest things to do together with your child.

Technical stuff gives you the real deal, mathematically speaking. Most of the time, you and your child don't really *need* to know the things that go with this icon, but sometimes you want to know the full story.

Beyond the Book

In addition to the content of this book, you can access some related material online. A series of videos that cover some techniques and big ideas from the book is available at www.dummies.com/commoncoremathvids.

Check out the related videos for additional help:

- ✔ Chapter 5: Making tens
- ✔ Chapter 6: Decomposing numbers
- ✔ Chapter 7: Eyeing adding and subtracting strategies
- ✔ Chapter 8: Using addition algorithms
- ✔ Chapter 9: Comparing fractions
- ✔ Chapter 10: Tackling multiplication algorithms

You can access a free Cheat Sheet at www.dummies.com/cheatsheet/commoncoremathforparents that contains additional information about the standards. You can also access some additional helpful bits of information at www.dummies.com/extras/commoncoremathforparents.

Where to Go from Here

Feel free to start reading from Chapter 1 to get an overview of what the book has to offer. You also can turn to the grade that interests you most, which may be the grade one of your children is in right now or (if you're a teacher) it may be the grade you teach. That grade most likely refers to another grade for more information. You can follow the references that interest you and skip the ones that don't. If you've been frustrated by a strange-looking homework assignment, get yourself to Chapter 4 to get an overview of the nature and purpose of homework in Common Core classrooms. Or you can flip through the table of contents or index to search for any topic that interests you.

Part I

Getting Started with Common Core Math Standards

Go to www.dummies.com/cheatsheet/commoncoremathforparents for a Cheat Sheet that gives you some easy-to-refer-to tips that can help you when trying to familiarize yourself with the Common Core State Standards for math.

In this part ...

- ✔ Understand how the Common Core State Standards fit in the history of math teaching in the United States and how math education has evolved during the last century.

- ✔ Look at the different ways that students are doing math in Common Core classrooms so that you know what to expect when your child enters a certain grade.

- ✔ Get tips about your child's homework so that you're better prepared to help and can reduce any related stress.

- ✔ Comprehend the purpose of some nontraditional homework assignments that you may have seen on social media (or in your own child's backpack!) to avoid frustration on your part or your child's.

Chapter 1

The Lowdown on Common Core Math, Just the Basics

In This Chapter

▶ Knowing what Common Core Math means

▶ Getting tips on helping with homework

▶ Developing math from kindergarten through high school

*I*n recent years, news outlets have regularly covered stories on the math that students are learning in school. Whether the story is about international comparisons of student learning ("You must panic! The United States is falling behind!") or the homework students bring home ("You must panic! Second graders are using number lines!"), these news stories have an element of urgency to them.

This urgency is understandable. Parents want their children to have the best possible opportunities in life and career. In a modern, technology-dependent society, a solid math background is an important part of creating those opportunities. People who struggle to work with numbers, spatial relationships, and algebra can't find employment in sectors that rely on technology and science, and more industries than ever do rely on technology and science.

You can think beyond the employment picture and still be concerned with how your child learns math. Everyday life requires more thinking about quantities than in the past. Is this week's cold weather evidence against global warming? Should I have my child vaccinated? What does it mean for my state's budget if everyone buys more stuff online? To answer these questions confidently requires more comfort with numbers than you need to count change correctly — which may have been a primary concern for citizens 100 years ago. You still need to count change correctly (or risk getting swindled on a daily basis!), but you need so much more than that to participate fully in the modern-day United States.

As of this writing, in 44 states and the District of Columbia — together totaling about 84 percent of the US population — have enacted the Common Core State Standards. Just like your child will need more math for career and citizenship than your grandparents needed, you need a bit more math than your grandparents did to understand what your child is doing in school. This chapter serves as your jumping-off point into the world of Common Core Math.

Understanding What Common Core Math Is

There really is no such thing as Common Core Math. Okay, you're scratching your head, so allow me to explain what I mean and why this book is so important.

In a Common Core classroom, students' ideas are center stage with the focus not on Common Core Math, but on student thinking. Teachers work every day to help students improve their thinking and to provide students with new ideas when they need them and when they're ready for them.

The Common Core Standards still require students to memorize addition and multiplication facts. They still require students to learn the standard algorithms and the Pythagorean theorem. None of those things have disappeared from the math curriculum. Instead, the role of student thinking has changed. Students' ideas are an important beginning place for math learning rather than being seen as an irrelevant distraction.

Many people in this country have experiences with school math that can be summarized as *rules without reasons*. They were told to *do this* in situation A, but *do that* in situation B. They never understood why and they struggled to remember whether to do *this* or *that* in situation A. And they struggled to tell situation A from situation B so they just applied what they hoped was the right rule in the right situation and prayed that they could earn enough partial credit to pass the test.

A quick story helps to illustrate. My mother-in-law, Lucie, is a fabulous woman. She wouldn't describe herself as a math person. While talking to her about math teaching (no one escapes that fate in my personal life), I asked her to calculate 1,001 – 2. She thought for a moment and said 999. I asked her how she knew, and she said that she had learned it in school. I didn't believe that for a moment — there is no way this particular fact was one that she had to memorize in second grade, plus I could see that she thought for a moment before responding. When I pressed, she finally was able to say that she knew 1,000 was one less than 1,001, and so 999 was two less than 1,001.

We talked about her solution, and she noticed that she had done something different in her head than she would have done on paper. The way she solved 1,001 – 2 was different from the way she was taught in school. For Lucie — and for far too many students — the methods taught in school are disconnected from the ways she thinks about numbers.

Lucie's way of finding 1,001 – 2 wasn't Common Core Math. It was just good mathematical thinking. The standard algorithm (see Chapter 10) is a correct but seriously inefficient way of finding 1,001 – 2. Similarly, it would be inefficient to use Lucie's strategy to find 1,001 – 999 (you would have to count backwards from 1,001 until you got to 2).

Examining the Standards for Mathematical Practice

One unique aspect of the Common Core State Standards is that their focus goes beyond the familiar content of numbers, geometry, algebra, and statistics. They also include a set of Standards for Mathematical Practice that describe how people work when they're doing math. These standards apply across all grade levels, with kindergarteners operating at a level of sophistication appropriate to them and high school students working at a much more sophisticated level.

The list of Standards for Mathematical Practice is fairly long — there are eight of them — and they overlap in ways that make it challenging for the average non-math teacher to tell them apart. But they're important aspects of the work that children do in Common Core classrooms, so in this book,

I have boiled the Standards for Mathematical Practice down to four simple statements about what students at all grade levels should be doing in math class. In Chapter 3, I describe these four statements in detail and relate them to the eight standards from the Common Core.

Ask questions

Students should ask questions such as, "What if . . . ?", "Why?" and "How do we know that?" They should also seek to answer these questions. These may not be the questions that you picture students asking in math class, but they're really useful questions for learning more math.

Play

When children play, they make things up and try out things. They don't worry about getting everything perfect. They repeat the same scenario many times, changing it a little bit each time to see what happens. They challenge themselves. They laugh.

All of this can happen in the math classroom, too. Math is challenging, but so are handstands, video games, and soccer. All of these activities involve risk-taking and exploration. Math should too. Often, the line between play and work is drawn with consequences. If an activity has high stakes, it isn't so much fun and turns into work. A Common Core classroom has many opportunities for students to play with math: to try something new, to create challenges for themselves and others, and to get things wrong and try again.

Math has right answers, just as football has touchdowns. But not every game is for the championship, and not every math activity needs to be high stakes.

Argue

Arguing is a highly mathematical activity. A good argument has some agreed-upon starting point, has some rules for moving forward, and seeks to uncover the truth. In a Common Core classroom, students have to figure some things out for themselves, which means that they need to formulate an argument to support their thinking. The sophistication of these arguments increases as students age and as they gain more practice.

For example, in second grade, a student might need to convince someone else that 14 is an even number. In high school, a student might need to write a proof that the sum of the angle measures of any quadrilateral is 360°. Both of these activities count as arguing.

Connect

Math is often taught as a long list of disconnected facts, but it shouldn't be. Mathematical ideas are connected to each other, and they're easier to use and to remember when students see connections among them.

Even memorizing multiplication facts — an activity that should be rich with connections — sometimes boils down to a conditioned response activity where each fact is different from each one. Math isn't memorizing. To be sure, being able to quickly remember multiplication and addition facts is useful, but an overemphasis on memorization can keep many students from developing the even more useful skill of thinking through things they don't know right away.

In a Common Core classroom, students spend time noticing how new ideas relate to old ones, how solutions to certain kinds of problems are just like solutions to others that seem unrelated, and so on. Looking for and talking about connections is an important part of doing math.

Looking at the Standards in the Different Grades

The Common Core State Standards have a grade-by-grade structure. This structure is required of state standards in the era of No Child Left Behind — the federal law that requires standards and high-stakes testing at the state level in return for federal education funding.

Several topics run through the grade levels, such as *geometry*, *the number system*, and so on. These topics are called *domains* in Common Core. The emphasis on these domains shifts through the grade levels — for instance, kindergarteners spend a lot more time with number and high school students spend a lot more time with algebra — but no domain is specific only to one grade level.

Understanding the ways the math builds from one grade to the next is important in knowing how to help your child stay focused on what is important. In the following sections, you find very brief overviews of each grade, and a reference to the chapter where I write about these ideas in depth. The chapter number corresponds to the typical age at each grade K–high school. So kindergarten is Chapter 5, first grade is Chapter 6, and so on.

Kindergarten

Kindergarten is a time of play, fun, and discovery. Kindergarten math is no exception. In kindergarten, students play with numbers and shapes. They count out loud — by ones and by tens — in order to get familiar with the word patterns in English number language. They count small quantities of objects and come to understand (if they don't already) that the last number you say answers the question "How many?"

They play with addition and subtraction by thinking about a variety of situations where small collections of objects are getting bigger or smaller, or where they are being compared. Kindergarteners play with shapes — naming them, sorting them, cutting them apart, and putting them back together. Chapter 5 goes in-depth on the kindergarten curriculum.

First grade

First grade is about addition and subtraction. Students work with one-digit and small two-digit numbers as they explore the relationships between addition and subtraction. Together, addition and subtraction are referred to as *operations*. Operations are the building blocks of algebra in later grades, so in a sense, first graders are studying algebra when they think about how $3 + 4 = 7$ relates to $7 - 3 = 4$ and to $7 - 4 = 3$.

In Chapter 6, I describe the full scope of first-grade math, and I also discuss why it's important for children at all grade levels to think their way through word problems rather than hunt for keywords.

Second grade

Second grade is about place value. (*Place value* refers to the idea that 12 and 21 — despite having the same digits — are very different numbers; the 2 in 12 is in the ones place, while the 2 in 21 is in the tens place — those 2s have different values because of the place they are in). This one topic is a stumbling block for students going through elementary school and into algebra if they don't make sense of it early in their math studies. In Chapter 7, I describe the challenges students encounter in learning place value, and how so much of the math that comes later depends on it.

Third grade

Third grade is about multiplication and division. In Chapter 8, I describe how multiplication and division have the same relationship to each other that addition and subtraction do. Multiplication tells you how many things are in some number of equal-sized groups, and division helps you figure out either the number of groups or the number of things in each group.

Third graders also study fractions in depth for the first time. In particular, they concern themselves with *unit fractions* — fractions that have a numerator of 1.

Fourth grade

In fourth grade, students use their knowledge of multiplication facts to do interesting things. They study factors and multiples. They develop techniques for multiplying large numbers together — both on paper and in their heads. They compare fractions and write equivalent fractions. All of this work depends on knowing single-digit multiplication facts. In Chapter 9, I explain the importance of multiplication across grade levels as well as describe the full scope of math in fourth grade.

Fourth-grade math has a good deal of measurement. Students look at relationships between units, such as feet and inches or meters and kilometers, and they study angle measurement for the first time.

Fifth grade

In fifth grade, students begin to dip their toes into algebra. They write simple expressions using variables to represent quantities they're trying to figure out, and they continue to deepen their understanding of the relationships among addition, subtraction, multiplication, and division. Fifth graders begin to add and subtract fractions.

In Chapter 10, I detail the work students do in all topics in fifth grade, and I describe the role and importance of standard algorithms in elementary school math.

The standard algorithms for addition, subtraction, multiplication, and division — what you may think of as *the old-fashioned way* — are all specifically required in the Common Core State Standards, although students learn other ways of doing these things, too (especially so that they can compute mentally).

Sixth grade

In sixth grade, students transition from the addition and subtraction world of elementary school to the multiplication and division world of middle school and of algebra. In earlier grades, students learned *how* to multiply and divide, but in sixth grade, they develop multiplication as a way of beginning to think proportionally by studying ratios and rates. A *ratio* is a comparison of two numbers that depends on a multiplication relationship; a *rate* is a description of changing values that depends on a multiplication relationship.

Students do a bit more sophisticated work with algebraic expressions in sixth grade, including making coordinate graphs to describe how two variables relate to each other. They begin to study interesting area measurement situations, mostly based on the area of a rectangle.

In Chapter 11, I describe all of this in greater detail, and I give you a crash course in dividing fractions (and this turns out to be surprisingly fun — go have a look!).

Seventh grade

Seventh grade is about deeply understanding proportional relationships and solving proportions. A *proportional relationship* is one where if you double one measurement, the other one doubles too. The relationship between feet and inches is proportional, for example. Three feet is 36 inches, while 6 feet is the same as 72 inches — both numbers doubled. Students notice proportional relationships in problems about rates of motion, unit prices, and circumference of a circle. They also notice when a relationship isn't proportional.

In Chapter 12, I write about how seventh graders also study operations on negative numbers, including answering the age-old question: "Why is a negative times a negative equal to a positive?" (See Chapter 12 for the answer to this question.) Seventh graders deepen their study of measurement by considering area and circumference of a circle, and they work hard to understand what information they need before they can conclude that two triangles are the same as each other.

Eighth grade

In eighth grade, students study algebra. Eighth grade algebra in the Common Core State Standards is the algebra of linear relationships. A *linear relationship* is one with a constant rate of change. Put another way, a linear relationship is one that has a straight line graph.

In Chapter 13, I describe the eighth grade standards in depth, including how a skeptical student is likely to think about irrational numbers (these are numbers, like π or the square root of 2, that you can't write as a ratio of whole numbers). Eighth graders learn about exponents and functions, and they solve systems of equations.

High school

In Chapter 14, I give you a whirlwind tour of high school math, including the different structures that schools and districts may choose for high school programs. Of special importance in this chapter is that the phrase *college and career ready* has implications in high school programs. High school students study math that will be useful to them as they move beyond the high school classroom.

Helping Your Child with Homework

Whatever the standards may be — Common Core or anything else — most children will be frustrated with a homework assignment from time to time. The advice to you, the parent, doesn't change just because your home state has adopted the Common Core.

What may change is the ways children are expected to work on their homework (refer to the earlier section, "Examining the Standards for Mathematical Practice" in this chapter for more details). Teachers may ask their students to practice something they worked on in class, but it may look unfamiliar to you. Productive ways of helping children with their homework are the same, though.

Don't do the homework for your child. Help your child clarify her thinking and identify what she knows and doesn't know. Monitor the difficulty level to make sure your child has interesting and challenging work, but not work that is far beyond her present abilities. Keep in touch with her teacher if things are out of balance so that you can work together for your child's benefit.

In the heat of the moment, though, you can easily lose sight of the big picture. So here are three simple tips for productive involvement with math homework. See Chapter 4 for more tips as well as some information about specific homework assignments that have become famous on social media:

- ✔ **Ask "How do you know?"** This question forces students to think about their own thinking, which is an important part of making that thinking better.

- ✔ **Wait for a response.** Ten or 15 seconds seems like a long time to sit silently when *you* know the answer, but it's not long at all to the person trying to figure the answer out.

- ✔ **Share a strategy.** After your child explains her thinking, talk about your own. If you want your child to engage with math homework, you can model it by talking about how *you* think about these problems.

Chapter 2

Looking at Math Teaching Then and Now

In This Chapter

▶ Comprehending the history of reform in math education

▶ Understanding the role of testing and accountability in the development of the Common Core

*T*he Common Core State Standards for mathematics are a sequence of math concepts and skills that students should learn in kindergarten through 12th grade. But these standards didn't just drop out of the sky like the tablets upon which the Ten Commandments were written. These standards have a history. Knowing a little bit about this history can help you understand why the standards are as they are and exactly what problems the standards are purported to solve.

In this chapter, I give you a whirlwind tour of the last 100 years (or so) of math education in the United States so you can grasp why Common Core is here today.

Setting Goals for the 1900s

In 1892, the Committee of Ten met with the charge of making reform recommendations for the structure and content of secondary education in the United States. At the time, Germany, Britain, and France were seen as the major international competitors against which to compare US schools and achievement.

This question of international competitiveness is frequently the concern in education reform initiatives (see the sections "Competing Globally with Advanced Math and Science" and

"Reaching Consensus with the Common Core" later in this chapter for specific reforms).

This committee, established by the National Education Association (NEA), put in place a structure for the high school curriculum that continues to predominate today — algebra in ninth grade, geometry in tenth grade, algebra 2 in 11th grade, and trigonometry or more advanced algebra in 12th grade.

With the exception of some initiatives that have sought to place eighth graders in algebra, some districts that view calculus in 12th grade as an important goal and some other tinkering around the edges, very little has changed about the structure of high school mathematics in the 130 years since the Committee of Ten produced its report.

Nonetheless, education reform initiatives have continued, and the question of international competitiveness is frequently the concern. Only the players have changed. The basic game of comparing the United States internationally to other systems of education remains the same.

Competing Globally with Advanced Math and Science

In 1957, the Soviet Union launched Sputnik — the first man-made satellite. This launch was a symbolic victory for the Soviet Union. It also was an event that sparked a great deal of concern about whether the United States had the scientific and mathematical infrastructure to be competitive in the so-called Space Race. As with both earlier reforms and later ones, international comparison was the catalyst for action.

One of the primary concerns was that the United States wasn't adequately preparing a generation of advanced research scientists and mathematicians for the work of outpacing the Soviets in space and defense. As is often the case, the nation looked to the public schools to solve this societal problem.

As one part of the effort to address the perceived lack of math and science research capacity, the National Science Foundation funded curriculum writing in mathematics — most notably through the School Mathematics Study Group. The writers on this project were primarily research mathematicians at institutions such as Yale University and Stanford University. These mathematicians set out to write the math curriculum they wished they had been offered as K–12 students. The resulting texts focused on the abstract foundations of mathematics — set theory, functions, formal logic, and so on. The collection of ideas in these texts became popularly known as *New Math.*

The important thing to know about the New Math textbooks is that their purpose was to better prepare the students who were supposed to go on to be the nation's mathematical and scientific elite. By contrast, later reforms (see the section "Rethinking Math Teaching in the Age of Information" in this chapter) were based on an interest in raising the mathematical achievement of the general population of students.

Eventually, the New Math reforms failed as a result of public distaste for the unfamiliar ideas coupled with a teaching corps that hadn't been adequately prepared to teach these ideas. Whether the programs would have been effective in an ideal situation is unclear, but in the messy world of American schools, they were a failure. The School Mathematics Study Group closed up shop in the 1970s, and by this time, few of the textbooks were still in use in the United States.

Returning to Basics

The 1970s were characterized by a call to get back to basics. The prevailing mood was that earlier reform efforts that had focused on preparing advanced mathematicians and research scientists had left out large numbers of students by inadequately preparing them in arithmetic and basic algebra. The feeling was that students had been taught the foundations of mathematics, but few of them could actually *do* any mathematics.

The popular argument of the time was that mathematics is a subject that builds from simple concepts to more advanced ones in a linear way and that the press to focus on advanced mathematical ideas throughout the public school curriculum rejected this basic truth. Step-by-step development of skills, with mastery at each stage before moving on to the next one, became the rule of the day.

Extreme versions of the back-to-basics perspective presented math as a disconnected series of facts. The job of teaching math — in this perspective — was to elicit the correct response to each stimulus. *Two plus three* should make a child say *five*, and *three plus five* should make a child say *eight*.

The 1970s was the era of timed math tests, a practice that is still common in American math education. In a timed math test, students may have had to write all of the multiplication facts involving 9 in one minute with the facts out of order — so 9 × 2, 9 × 7, 9 × 1, and so on. Students who mastered their nines times tables this way would move on to the tens times tables and so on. The key to a timed test is that it doesn't allow time to think: The goal is memorization.

Later reforms were based on different ideas about how children learn mathematics, as I discuss in the next section.

Teaching Math in the Info Age

In 1989, after quite a few years of discussion and work, the National Council of Teachers of Mathematics (NCTM) released the *Curriculum and Evaluation Standards for School Mathematics*, commonly referred to as the *NCTM Standards*. This document was controversial from the get-go, but it was notable for two reasons:

- ✔ It presented a comprehensive view of the contemporary and future landscape of math teaching.
- ✔ A professional organization of teachers, rather than a government entity, produced it.

A prominent feature of the NCTM Standards was a call for reconsidering the role of mathematics education in the era of computational technology. By 1989 people had been able to carry around calculators and had access to them for simple daily living tasks, such as balancing a checkbook, for a number of years. Also, the committee clearly could see that access to computational power would be greatly increasing in the ensuing years, which has certainly proven to be the case.

The NCTM Standards built the case that the mathematics taught in schools ought to be reconsidered in light of this ready availability of computing power. Much of the mathematics that students studied in school at the time was focused on training students to compute. The NCTM Standards offered a vision for how this should change in an era where computers would crunch numbers with unprecedented speed and economy.

The NCTM Standards laid out proposals for these changes in detail. They called for less paper-and-pencil computation, and more estimation and problem solving.

Many members of the public interpreted (rightly or wrongly) these changes to be calls to do away with teaching students computational skills. As with earlier reform efforts, this public pushback led to controversy and conflict. This time, the conflict had a name — the Math Wars.

The Math Wars were focused in particular on curriculum projects developed through funding in the 1990s by the National Science Foundation. Certain groups of mathematicians and scientists saw in these curricula an overemphasis on student-developed methods and too many traditional topics being left out. Over time, the Math Wars settled down with both sides tempering their language and tactics. Most of the curricula in question persisted in some form beyond the Math Wars, and several are still in common use today.

One important feature of the NCTM Standards was that math topics were arranged by grade bands rather than by individual grades. The standards called for children to learn something in sixth grade through eighth grade, but didn't specify which grade. This allowed for some diversity in curricular approaches. One set of materials could teach operations with integers (as I discuss in Chapter 12) in sixth grade, for example, while another could teach it in seventh grade. To be aligned with NCTM Standards, all that mattered was that this topic appeared somewhere in middle school.

Holding Teachers Accountable

In 2001, as part of a larger reform effort intended to address widely varying local standards, the achievement gap between white students and students of color, and a host of other concerns including adequate education for students with disabilities, the No Child Left Behind Act became law.

No Child Left Behind (NCLB) mandated for the first time

- ✔ Year-by-year standards in reading and math at the state level
- ✔ Year-by-year testing that was to be standardized within each state
- ✔ Progress toward the goal of 100 percent student proficiency by 2014
- ✔ An escalating series of consequences for schools that fail to make progress toward this goal
- ✔ Disaggregation of scores for students based on home language, ethnicity, disabilities, and so on

People disagree greatly about whether NCLB has been a net benefit to the public school system, but the program certainly has had both benefits and costs. Many classroom teachers (and parents) argue that the increased scrutiny and pressure for improved test scores have been detrimental to creating meaningful school experiences; preparation for annual testing has become a priority. This claim is especially true for subjects besides reading and math, which are the only two subjects for which NCLB requires testing.

Also, 100 percent proficiency was an unrealistic goal. Although having goals and high standards is good, schools have been measured against progress toward that goal, which means that excellent schools with initial high proficiency rates were being penalized and labeled as _failing_ when they were unable to move closer to the 100 percent proficiency mark.

On the other hand, disaggregating test score data has certainly been helpful for schools to better understand how they're serving the needs of various populations. Advocates for learning disabled students, for example, have enjoyed having better data to understand the relative success of these students in schools. Similarly, schools can better see whether minority populations are being served as well as the majority population. This information would be difficult to know without disaggregated score reporting as mandated in NCLB.

One concern developed in the years after the passage of NCLB: Each state wrote its own standards, which could vary widely. It had consequences for textbook publishing that, in turn, impacted students in US classrooms. In the United States, textbook publishing has a major influence on how school subjects are taught in the classroom, and major textbook publishers have an incentive to craft textbooks that will meet the standards of as many states as possible.

On a practical level, this meant that after passage of NCLB and the writing of state standards, publishers needed to create textbooks that included each topic in all possible grade levels where it would be taught. The logic that continues to prevail in textbook publishing decisions today is that teachers can skip content that doesn't meet their state standards, but that districts won't buy materials that don't cover all required standards at each grade. In reality, decisions about what part of a textbook a teacher uses are far more complicated than these assumptions. Nonetheless, the need to sell the same book in all 50 states dictates the content that gets included.

The result is often described as a *mile-wide, inch-deep* curriculum. The *mile-wide* part refers to the fact that many, many topics are taught at each grade level in a typical US math classroom — many more than in a typical Japanese or German classroom, for example. The *inch-deep* part refers to the fact that American students are much less likely than their Japanese or German counterparts to spend enough time with any one topic to know it well. Since they don't know it well, they review it in the next grade, which widens the curriculum in that grade, and so on in a perpetual cycle.

Agreeing on the Common Core

In 2009, the National Governors Association (NGA) — together with the Council of Chief State School Officers (CCSSO) — convened a committee to write, solicit feedback, revise, and produce the Common Core State Standards (CCSS) for mathematics and English language arts. The goal was to create a set of standards that would prepare students for college and the workforce and to create national consensus about what this preparation looks like — down to the details of topics to be studied at individual grades.

The standards were released in 2010. Soon 44 states and the District of Columbia adopted the standards. Many of these states adopted the standards prior to completed versions of the standards being released, and many did so in order to compete for federal funding under a program called *Race to the Top.*

A spirited and principled debate can be had about the Race to the Top program, and whether it's an appropriate use of the powers of the Department of Education. But the Common Core State Standards didn't arise from this program, and they aren't the law of the land at the federal level.

A founding principle of the Common Core Standards is that of international benchmarking. That is, the authors of the standards were charged with comparing their work against the standards of high-achieving countries such as Singapore, Japan, and Germany. The idea that the intertwined nature of world economics requires American education to be attentive to international standards of education is a very old one.

One of the hopes for the Common Core Math Standards is that they'll allow for better, more coherent curriculum materials being produced by major publishers for classroom use, thus helping to eliminate the mile-wide–inch-deep dilemma created by NCLB. With the Common Core, most states now require the same set of topics at each grade level. As a result, the focus for developing published classroom materials should shift from trying to cover every possible topic in all 50 states to covering smaller, more focused collections of topics well.

The rest of this book deals with the specifics of the Common Core State Standards for Mathematics. Looking back at the reforms of this chapter, it seems unlikely that Common Core is going to be the last and one true solution to American society's fretting about children's math learning. But these standards are in place in the vast majority of states right now. Knowing a little bit about how these standards came to be will be helpful as you read other parts of the book and come to know how students are learning in classrooms today.

Chapter 3

Exploring the Standards for Mathematical Practice

In This Chapter

▶ Getting a different perspective on asking questions in math class

▶ Seeing the importance of arguing in math class

▶ Playing with math

▶ Making connections

*T*he Common Core State Standards for Mathematics is more than a document that spells out the necessary content for grades K–12. It also contains the *Standards for Mathematical Practice (SMP)*. These standards describe the activities that are involved when people do mathematics — whether these people are students, parents, or mathematicians. They're the ways of thinking and of working that make math different from (say) art, English, philosophy, or science.

The SMP look different at different grades, as students gain more practice and as the content becomes more sophisticated. But the basic set of practices applies in some form at every level.

In this chapter, I touch on each of the eight SMP, but not in the order or format that they appear in the Common Core State Standards. I don't because that list is overwhelming and the ideas overlap in ways that make it difficult to follow unless you spend every waking moment thinking about them. As a result, I have abbreviated this list — to four main concepts — which still retains the flavor and spirit of the SMP without the overwhelming length.

Focusing on Asking Questions

Asking questions is a huge part of learning. The SMP acknowledge and incorporate asking questions into instruction, but in a sense that may be different from how you imagine based on your own experience as a math student.

In a Common Core classroom, students should be asking all different types of questions of themselves and of others. The following sections take a closer look at these types of questions.

Asking "Why?" and "How do you know?"

Questions, such as "why" and "how do you know" (or "how do *I* know"), are very similar. They force you to build an argument. In fact, "Construct viable arguments and critique the reasoning of others" is one of the Common Core SMP.

One common claim about math in elementary school is that students need to *just know* their number facts, such as all the sums of two one-digit numbers ($3 + 8$, $7 + 5$, and so on). Students certainly do need to be able to produce single-digit sums quickly, and Common Core requires it in second grade (refer to Chapter 7).

You may not have stopped to consider why $8 + 4 = 12$ for quite some time. In fact, asking the question "Why is $8 + 4 = 12$?" may seem totally silly. You just know that it is.

But a slightly different version of the question may seem more compelling. How do you *know* that $8 + 4 = 12$? Or consider this version: If you couldn't remember what $8 + 4$ was for a moment, how could you figure it out?

Maybe you could count it out. Start at 8 and then count four more — 9, 10, 11, and 12. That counting ends on 12, so $8 + 4 = 12$. Although this strategy is a lot less practical for something such as $9 + 9$, and it doesn't transfer well at all to multi-digit addition, it's a viable strategy for explaining why 8 and 4 make 12.

Here's another way to know this number fact — one that is somewhere between counting and memorizing yet is powerful enough to work with many different kinds of numbers. Ten is an important number (check out Chapter 6 for more information). If you know that $8 + 2 = 10$ and you also know that 4 is two more than 2, then $8 + 4$ is two more than $8 + 2$, so it's two more than 10, which is 12.

When students relate number facts to each other and use ideas in addition to recalling facts, they're working with one of the Common Core SMP: "Looking for and making use of structure."

Teachers don't *tell* students these things in a Common Core classroom. Rather they emphasize the important things (such as that ten is an important number), give students lots of experience working with these important things and provide lots of feedback, and ask them "why" and "how do you know." Teaching certainly involves telling students some things, but the SMP call on students and teachers both to ask more questions.

Asking "What if?"

This question is at the heart of nearly every historical advancement in mathematics and should be at the heart of every student's learning. Think about negative numbers, for example. Nearly everyone has the experience of being in second grade and listening to the teacher explain how to subtract $35 - 19$ using the standard algorithm (see Chapter 10 for more on standard algorithms). "You can't take 9 from 5," the teacher says referring to the units place, "so you need to borrow from the tens place."

But what if you *could* take 9 from 5? What if $5 - 9$ *did* have an answer? Second graders sometimes ask similar important questions, and the answers to these questions involve negative numbers. A whole new kind of number comes from asking "what if?"

Similarly, fractions come from asking "What if you could share 5 things equally among 3 people?" and complex numbers (a high school topic, see Chapter 14 for more information) come from asking, "What if you could find a square root of a negative number?"

Asking "Is it good enough?"

"Attending to precision" is one of the eight Common Core SMP. Students attend to precision when they ask themselves, "Is it good enough?" This question is about numbers, but it's also about communication. At a basic level, precision means making judgments about measurements and calculations by answering questions such as these:

- When is it okay to estimate?
- When is it okay to use 3.14 for π?
- Should I round this number? To what place?
- Is $\frac{1}{3}$ really equal to 0.33?

The answers to these questions relate to being precise about the use of numbers, which is important in mathematics. Just as important, though — and often overlooked — is attending to precision in other areas of doing and communicating mathematics.

Being careful with units is an important part of doing mathematics, and it's part of *attending to precision*. Any time you express a quantity, a unit goes along with it. Two plus two equals four because two *things* and two more *things* total four *things*.

Paying attention to units is important when students are learning about multiplication in third grade (see Chapters 7 and 8 for more about units and multiplication) because 3 × 4 means *3 groups of 4*. The 3 and the 4 refer to different units. You can't multiply three eggs by four eggs and get anything meaningful. But you can multiply three cartons by 4 eggs per carton to get 12 eggs all together.

Paying attention to units is important when students are studying fractions in fifth grade because the algorithm for adding fractions depends on units (see Chapter 10): $\frac{1}{4} + \frac{2}{4} = \frac{3}{4}$ because *fourths* are units. You have one-fourth and two more fourths for a total of three-fourths.

A final example of attending to precision is in the use of language. Mathematicians use words very precisely, and it begins with precise definitions. *A circle is a round shape* isn't a precise definition — it isn't good enough to sort out circles from all things that aren't circles. An oval is round but not a circle, for example. A precise definition for a circle is *the set of points in a plane that are the same distance from a common point, called the center.* In a Common Core classroom, the need for precise definitions should come from exploring ideas and arguing about the meaning of words. Questions such as "Is a square a rectangle?," "Is 0 even?," "Is 1 prime?," and "Is $y = 1$ a linear function?" all depend on precise definitions of the relevant words — *rectangle, even, prime,* and *linear.*

Attention to the precision of a definition doesn't need to wait for upper elementary or middle school, though. Young children are ready to play with definitions even before they attend school. "Is a hot dog a sandwich?" is a lovely question accessible to young children that helps them attend to the precision of the meaning of *sandwich.* Is your definition of sandwich good enough to determine whether a hot dog is one? Making lunch can be a chance to work on mathematical practices!

Asking "Does this make sense?"

Students need to make sense of problems when they do math. The meaning of *a problem* varies widely from kindergarten through high school, but making sense of problems is important throughout. "Does this make sense?" is an important question for students to ask themselves as they work. A Common Core SMP addresses this question: "Making sense of problems and persevering in solving them."

First graders may meet a problem such as this: Elle and Josh bought some apples, but Josh ate five while Elle was at work. Now there are seven left. How many did they buy?

A student who tries to solve that problem *without* making sense of it may search for so-called *keywords*, find the word *left*, remember that it means to subtract, and subtract five from seven to get two (see Chapter 6 for more about keywords). For a first grader, making sense of a problem means understanding what is going on, which may involve drawing a picture or telling a story that relates the problem to her own life. It does *not* mean just translating the problem into mathematical symbols. A first grader needs to ask herself: "Does it make sense to subtract these numbers?"

An eighth grader, by contrast, may have a problem based purely in mathematical relationships, such as "Does doubling the x-value in a linear function always double the y-value? Does it ever?" Making sense of this problem for an eighth grader requires drawing on more vocabulary (*linear, function,* and so on) than a first grader will likely require, and it involves a good deal more sophistication. But the idea is the same: *making sense of problems* is very different from *knowing what value to substitute and computing correctly.*

Asking whether something makes sense is part of *persevering.* When you ask the question, the answer may be *no,* and that may require trying again. At all grade levels, when students persevere in solving problems, they're willing to accept the fact that something they try may not work. They're ready to start over again and to spend time trying to figure out things. A classroom (and home) climate that values working hard and learning from mistakes is helpful in fostering perseverance in problem solving. A classroom climate that only values right answers is less likely to encourage students to persevere.

Asking "What's going on here?"

Structure refers to how things are put together in math and to how things relate to each other. Structure reveals what is going on beneath the surface, which is why a Common Core SMP is "Look for and make use of structure." You can make a habit of looking for structure by asking, "What's going on here?"

If you have \$12.00 and I have \$3.00, you could say that you have \$9.00 more than me. That is making use of the *additive structure* of numbers — $3 + 9 = 12$. But you could also say that you have *four times as much* as I do. In that case, you're using the *multiplicative structure* of numbers — $4 \times 3 = 12$. You can ask, "What is going on with this comparison?" and "Does it make more sense to compare by multiplying and dividing, or by adding and subtracting in this situation?" (Chapters 7 and 11 discuss the additive and multiplicative structures in greater depth.)

When students sort shapes, they can pay attention to what the shapes look like, or they can pay attention to the mathematical structure of the shapes, such as by the number of sides, relationships among the sides, or the measures of their angles. This transition from noticing that rectangles look like doors, for example, to noticing that rectangles have four right angles is a long process in elementary school that begins in kindergarten (refer to Chapter 5). The question "What is going on with these rectangles that makes them rectangles?" is important here.

Look at the sequence of shapes in Figure 3-1. Each shape is made of one or more square tiles. If the shapes were to continue growing in the same way these do, how many tiles would be in the fifth shape? How many would be in the tenth shape? How many would be in the 100th shape?

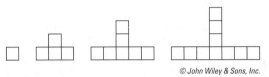

© *John Wiley & Sons, Inc.*

Figure 3-1: From left to right, the first through fourth shapes in a sequence.

Using the structure of these shapes, you may say that each shape consists of a tile in the center (shaded in Figure 3-2), surrounded by three arms, each with the same number of tiles. If the number of the shape is n, the length of each of these arms is $n - 1$, so the total number of tiles is $3(n - 1) + 1$.

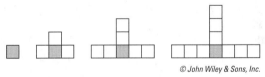

© *John Wiley & Sons, Inc.*

Figure 3-2: One way to see the structure of the shapes.

Other people may see different structures. The point is that seeing the structure of these shapes allows you to calculate the number of tiles in any of these shapes. The 100th shape has 298 tiles, for example.

Looking for and making use of structure is what students do when they do mental math. Something as simple as multiplying 4×10 by thinking about it as *four tens* or *forty* is an early place-value example of structure (see Chapter 7 for more on the structure of place value).

A final example of looking for structure is noticing the big picture of what algebraic symbols are telling you. If you know that n is a whole number, for example, then $2n$ is an even number and $2n + 1$ is an odd number (see Chapter 10 for more about even and odd numbers, and their algebraic representations). What's more, the expression $2n + 1$ represents *all* odd numbers. This is the power of looking for and making use of structure — representing infinitely many things in a single short expression.

Playing with Math

Ideas are things you can play with, and math is about ideas.

When children make up stories, parents and teachers are usually delighted to listen. Children are brought up with books and stories, and watching them practice writing their own can be exciting. When children make up games, or build things, or start playing instruments, parents and teachers can see the connection to playing.

But with math, people's vision can get clouded. It's easy to be concerned with *right* and *wrong* in math, while losing sight of the importance of playing with ideas. Yet playing is just as possible and just as useful in math as it is in other areas of children's lives. These sections look at some examples of math play, as they may happen in classrooms and at home.

Experimenting with symbols

One of the Common Core SMP is "Reason abstractly and quantitatively." This mouthful refers to the idea that mathematical symbols have meaning, and that students should strive to keep that meaning in mind as they work. Okay, I realize that it doesn't sound very playful, so stick with me here.

For example, a class full of second graders can discuss how many muffins are in the partially filled muffin tin in Figure 3-3. Some of these second graders will count the muffins one by one. Others will see more sophisticated relationships. The teacher may help students record their thinking using arithmetic.

Can you see how each of these expressions correctly counts the muffins? What could the child who wrote this expression have been thinking?

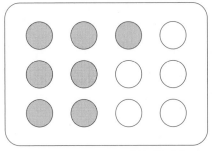

Figure 3-3: How many muffins? How did you count them?

- $3 + 3 + 1$
- $12 - 3 - 2$
- $\frac{1}{2}$ of $12 + 1$
- $2 + 2 + 3$
- $3 \times 3 - 2$

When you (or a second grader) are working on a task such as this one, you're playing with math. You're imagining new relationships, trying something without fear of getting it wrong, and seeing a small corner of the world in new ways. Children do all these things when they play.

There is no *one answer* involved when you ask, "How can you count these muffins?" — all of the ways of counting come up with seven muffins. That leaves you free to play with the ideas and see whether you can see what someone else sees in this situation.

As far as the Common Core SMP is concerned, you probably found yourself going back and forth between the symbols (3×3) and the meaning (Where can I find 3 groups of 3 in this picture?). This is the heart of "Reasoning abstractly and quantitatively" — moving between the context (counting the muffins) and the abstract ideas $(3 \times 3 - 2)$.

Building models

Building is a form of play. Building with blocks, building sand castles, building a catapult to launch a pumpkin across a farm field, and building a mathematical model are all different forms of the same playful human instinct to create.

When children make predictions, they're building mathematical models. For example, my son's bedtime went from 7:30 to 8:00 when he turned 8 years old. He took this to be a rule and began predicting future bedtimes at future ages, assuming that his bedtime would advance by a half hour each year. He extrapolated in both directions to determine that he would go to bed at 2:30 in the morning when he turns 21 and that he must have gone to bed at 6:00 when he was 4 years old. He built a model of the relationship between his bedtime and his age, and he played with that relationship.

Similar things happen in Common Core classrooms. Students identify patterns, they make predictions based on these patterns, and they check their predictions against reality. The Common Core SMP is "Model with mathematics."

Arguing Is a Ton of Fun

An *argument* in math isn't really the same as an argument your children may have in the back seat of the car. The back seat argument is likely to devolve quickly into name-calling, spitballs, and fisticuffs if you as a parent don't intervene. This behavior shouldn't happen in a mathematical argument. What these two kinds of arguments have in common is disagreement. In math, the disagreement is usually around whether something is true.

This SMP may be the most important: "Construct viable arguments and critique the reasoning of others." That's because argumentation is at the heart of math as a discipline.

Although many people perceive that math is about computing with numbers and variables, it's really about making particular kinds of arguments. Mathematicians base their arguments on things they already know are true and on the rules of logic. If something is true in math, you can know it in a much more certain and timeless way than in any other subject.

You can help your child practice making arguments by asking one simple question on a regular basis: "How do you know?" (Refer to the earlier section, "Asking 'Why' and 'How do you know?'") If your 7-year-old says that she ate four pistachios because she has eight pistachio shells on her plate, ask her how she knows eight shells mean four pistachios. If your 9-year-old claims that $1\frac{1}{2}$ scoops of $\frac{1}{2}$ cup of flour is $\frac{3}{4}$ cup, ask him how he knows. Make a habit of asking your children how they know so that they can start making a habit of building mathematical arguments.

If you have ever tried to explain to a child exactly why he can't do something, you have likely noticed that children are very good at critiquing the reasoning of others. Even toddlers can be like little lawyers when there is candy or some other privilege or liberty at stake, which can be tremendously frustrating as a parent, but it's a highly useful mathematical technique. Encourage it when you can and continue to lay down the law when you need to.

Connecting Ideas

Remembering a set of disconnected facts is more difficult than remembering a story. The typical human brain is very good at telling and remembering stories, while also very bad at retaining lists. A major difference between a list and a story is connections.

Fortunately (and not coincidentally), math is rich with connections. *Connections* basically are relationships between ideas. But math is often taught in a way that obscures these connections. When teachers insist on memorizing both addition and subtraction facts (which is different from insisting that students be able to produce these facts quickly), they obscure the connections between addition and subtraction, for example. If you know that $8 + 4 = 12$ and you know the connections between addition and subtraction, then you can quickly produce $4 + 8$, $12 - 4$, and $12 - 8$ because they're all connected examples of one relationship.

In fact, mathematics can connect what seem to be very different situations. When you build a mathematical model of a situation, you strip away details. For example, when you have three red apples and four green apples and you write $3 + 4$, you strip away the fact that this is an idea about apples. Then, when you count four spaces to the right on the number line, starting at 3, and you write $3 + 4$ again, you have a connection. This relationship — that $3 + 4 = 7$ — is true of a wide variety of contexts.

The following two questions are common in algebra and geometry classes:

- ✔ If n people are on a basketball team and each one gives the others one high five, how many high fives are given altogether?
- ✔ How many diagonals can you draw in a regular polygon with n sides?

In both cases, the answer is the same: $\frac{n(n-1)}{2}$. After you have enough experience with these types of problems, you can start to have hunches about which kinds of mathematical models are likely to be useful for different situations. Through repeated experiences with modeling, you can get better at noticing the structure of a problem situation.

In addition to thinking of algebra as *generalized arithmetic* (this means that algebra answers questions about *all* numbers, not just the numbers in a particular computation; refer to Chapter 10 for more information about generalized arithmetic and its role in algebra), you can think of algebra as an efficient way of getting things done. Algebra can capture the regularity in repeated reasoning. In order to capture it, you need to look for connections.

For example, when students study inverse functions in high school (as I discuss in Chapter 14), they can notice the connections between the inverses they find for linear functions. Namely, that the inverse of a linear function $y = mx + b$ is also linear and has a particular form: $y = \frac{1}{m}x - \frac{b}{m}$. This observation is motivated by repeatedly reasoning your way to a solution for particular linear functions and eventually asking the questions, "What is the same in each of these problems? How are these connected?"

A simpler example of this SMP about regularity in repeated reasoning occurs when students are moving from counting to solve problems such as $9 + 2$, $9 + 3$, and $9 + 4$ to having strategies for knowing the sum of 9 and any one-digit number. Students frequently notice that these sums come out as 10 plus a number one less than the original number. That is $9 + 2 = 11$, and 11 is $10 + 1$. Before students have memorized all of their single-digit addition facts, they frequently have noticed the regularity in their repeated reasoning about sums involving 9 — and similar patterns in sums involving other numbers.

Chapter 4

Understanding Homework Assignments

In This Chapter

▶ Going behind the scenes on homework

▶ Perusing some famous homework assignments

▶ Finding relief from homework frustration

*T*his chapter may get to the very heart of why you bought this book. A few of your child's homework assignments may have flummoxed you, and you want to know how to support your child. If so, you've come to the right place.

In this chapter, I give you a glimpse into the minds of teachers by summarizing the role teachers want homework to play in your child's education. (Spoiler alert: It's more complicated than you may think!) I also walk you through a few homework assignments that have become famous because they frustrate parents and give you some tips on supporting your child when he is struggling.

Getting the Inside Scoop: Teachers' Views on Homework

Knowing something about teachers' reasons for assigning homework can help you to know how to support your child at homework time. The issue of homework can be more complicated than you might expect.

Teachers assign homework for many reasons. Here is a partial list:

- ✔ To practice skills learned in class
- ✔ To introduce new ideas that there isn't time to cover in class
- ✔ To set the stage for the next day's lesson
- ✔ To teach responsibility
- ✔ To review things from previous grades that will be needed in the current grade
- ✔ To conform to school or district guidelines
- ✔ To conform to the expectations of parents and students
- ✔ To compel students to think about math outside of the math classroom
- ✔ To assess what students know in order to assign grades or to make decisions for future lessons

Most teachers can probably point to two or more of these reasons as their own. Many teachers can probably add another reason or two to the list.

Most teachers prefer to view themselves as partners with parents with the common interest of their students' success. To that end, you can ask at the beginning of the year (or anytime, really) what your child's teacher has as goals for homework and how he would like you to help out when your child gets stuck.

Most teachers and parents agree that a child's emotional and physical well-being needs to take precedence over any particular homework assignment. If your child occasionally struggles with homework, you may be able to let it go for a night or two as long as you make a plan with your child and stay in touch with the teacher. Write a note to the teacher explaining the situation and help your child catch up on necessary work over a few days or the weekend.

If you notice a pattern where your child is consistently struggling with homework, though, you'll want to sit down with the teacher to find out his purposes for homework and to develop strategies to meet those purposes.

Examining the Homework Assignments on Social Media

If you use social media such as Facebook or Twitter, you have probably seen at least one photograph of a math worksheet with comments suggesting that the nation is going downhill because of this new fuzzy math that kids are being taught.

This section looks at some of these specific worksheets to help you understand their intention (with an understanding that not everything designed with good intentions comes out well). I don't apologize for these examples, nor do I try to convince you that each is an excellent example of what today's educational system has to offer. I simply help you to understand what the teacher may have been hoping for in assigning them. And that is pretty much all any parent wants when faced with a ticking clock and a frustrated child at the end of a long day.

Refer to www.dummies.com/extras/commoncoremath forparents for another example of a troublesome homework assignment.

Number lines

One homework assignment seen by millions of people asked students to write a letter to a fictional student, Jack, who was trying to find the difference 427 – 116, and who drew a number line to keep track of his thinking. But Jack made a mistake. Students were supposed to find the mistake and write a letter to Jack explaining the error. Jack's number line looked something like Figure 4-1.

© John Wiley & Sons, Inc.

Figure 4-1: Explain this student's error in finding 427 – 116. _____

The reason millions of people saw this assignment was because a parent posted the following note to Facebook, addressing Jack.

Don't feel bad. I have a Bachelor of Science degree in Electrical Engineering which included extensive study in differential equations and other higher math applications. Even I cannot explain the Common Core Mathematics approach, nor get the answer correct.

The note shows the standard algorithm (which I explain in Chapter 10), suggests that this much simpler method would be preferred in industry, and is signed *Frustrated Parent.*

Frustrated Parent didn't intend to create a cultural phenomenon; he needed to let off a little steam at the end of a difficult night of getting his son through his assigned homework. With the Internet being what it is, though, this assignment was one of several used to argue that Common Core is bad for children. After all, if a parent who studied engineering and differential equations can't figure it out, how can a second grader?

This may not have been the right homework assignment for this child on this day. Nonetheless, here is what the assignment seems to have been going for:

- ✔ **Practice finding errors.** People make mistakes. Teachers constantly ask students to look for their own mistakes, and sometimes they ask students to find others' mistakes, which is similar to having students practice proofreading as they learn to become better writers. This assignment asked students to find someone else's error — to proofread another student's math work.

- ✔ **Use number lines.** The number line is a very useful tool in mathematics. (A *number line* is simply a line marked at regular intervals with each mark corresponding to a number. Depending on the grade level, a number line may have whole numbers, fractions, decimals, and/or negative numbers indicated. Numbers on a number line increase from left to right.) One way mathematicians describe the meaning of –1, for example, is by saying that it is one unit to the left of 0 on the number line. A thermometer is just a vertical number line (and negative numbers on that thermometer are very relevant to people like me who live in Minnesota). The coordinate plane where you graph lines and parabolas is made up of two number lines — one horizontal and one vertical. Your own elementary classroom when you were in first grade may have had a large number line running along the top of one wall of the classroom. The number line is an important way of visualizing numbers.

You may not want to solve a subtraction problem on a number line. And your child may not want to do that either, which is okay. But one way to think about subtraction is as the *distance on a number line* (refer to Chapter 12 for more details). That's what the fictional Jack was doing. Jack tried to find the point that is 116 units away from 427. That's a smart thing to do, and it's something the typical second grader can learn to understand.

But of course, not every second grader will be successful with finding another student's error represented on a number line. In fact, *many* of them won't be able to do this challenging task the first time that they see it. Refer to the later section, "Becoming Unstuck: Here's What You Can Do" for suggestions on how you and your child can proceed if you end up in the same situation as Frustrated Parent.

Dots

The New York Times published an article about the Common Core Standards that cited another frustrated parent. This time, the daughter was frustrated by "having to draw all those little tiny dots."

Here is an example of a second-grade task that may lead to students drawing dots:

> There are 8 cars in the school parking lot. How many wheels are there altogether?

A second grader isn't expected to know his multiplication facts. Yet, figuring out how many total wheels 8 cars have is reasonable. Many students will draw a picture, such as the one in Figure 4-2. Teachers encourage this kind of picture drawing because it helps students see the structure of the problem. A picture made of dots is much less tedious to draw than a picture of cars.

Drawing dozens of dots alone doesn't teach math. Instead, noticing and using the structure of the array of dots is where the learning takes place. When students study multiplication, they need to learn to see groupings. Multiplying 8 × 4 means finding out how many things are in eight groups of four. When teachers assign tasks such the one in Figure 4-2, they want

© John Wiley & Sons, Inc.

Figure 4-2: A child may draw to understand the problem.

students to practice finding products ($8 \times 4 = 32$), and they want to give students practice noticing the grouping that takes place when you multiply.

The goal of this kind of problem is to reinforce the idea of same-sized groups — an important building block for the later study of multiplication. The goal of an activity such as this is to help students focus on the meaning of multiplication before they learn their multiplication facts. If drawing dots is tedious for your child, you can achieve the same effect using pennies, paper clips, or any other small objects that are found in large quantities in a typical home. It's the same-sized groups that matter here, not the dots themselves.

Becoming Unstuck: What to Do

A few of the more stressful parenting moments include when time is tight and when your child is struggling with a homework assignment. Instead of cursing the heavens, Pythagoras, and the originators of homework, you need a strategy. Here is a good one, in three parts:

✔ **Listen.** Ask your child to tell you what he is supposed to do. If something needs to be done a particular way, listen to what it is. Ask what he knows and doesn't know and then listen to the response. Don't try to tell him what to do or show him the way that you remember from grade school. Just listen.

✔ **Help.** After you have listened, decide how you can help. Sometimes you may have to show your child the way you learned to do something. Sometimes you may have to look something up online together (see Chapter 15 for some possible resources). Sometimes you may have to phone a friend — your child's friend who may have an idea about what to do or your friend who may have some expertise. Don't assume that the way you learned is the *best* way (nor that it is the *worst*). An alternate way of working through something may help a child understand the meaning of what he's doing.

✔ **Let it go.** You and your child shouldn't be expected to spend your evenings engaged in tears and combat in the name of multiplication facts. As a parent, model good study habits by establishing that some time will be spent doing what is assigned. You also should model good mental health habits by setting limits on that time. When your child is no longer making productive progress, let this homework go. You can write a note to the teacher explaining the decision that you have made and get on with the rest of your evening.

✔ If getting stuck is a rare or occasional thing, letting it go is a good strategy to practice. If your child is regularly getting stuck on his homework and you find yourself letting it go on a regular basis, you probably need to have a conference with your child's teacher so that you each understand the demands being placed on your child. Frequent homework struggles can be a sign that your child needs something different from what he's getting at school. In such a case, you and your child's teacher should figure out how to work together to get him what he needs.

Helping Your Child without Doing the Work Yourself

Researchers have found that the human mind is surprisingly lazy. If it can get something done without thinking hard, it will. This research is no surprise to someone who has witnessed a typical 8-year-old child eager to get through her math homework so that she can go outside and play. "Is this right?" she asks, hopefully. "Is it right now? What about this?" This routine can quickly devolve into a guessing game rather than a chance to learn anything useful.

Furthermore, an important function of homework is what teachers call *formative assessment*, which means learning what students understand while they're still studying it. A classroom full of correct homework papers can signal to a teacher that everyone understands and it's time to move on. Now imagine if those correct homework papers were the result of parents (or tutors, or older siblings) walking these children through the steps of the homework problems. It could mislead the teacher. It's much better to write a note explaining what the student is struggling with (or to let your child submit some incorrect or incomplete problems) than to let the teacher think that your child understands something that he doesn't.

Parents and teachers can try a couple of useful strategies for getting students to think instead of guess and so to learn something rather than just stumble onto a right answer. Make any or all of these a part of your regular homework routine and you may find your child getting better at thinking for himself:

✔ **Ask "How do you know?"** Ask this question frequently. Question right and wrong answers. This question has many variations. "How do you know this is right?," "How did you know that was wrong?," "How did you know to do that?," and so on. This type of question forces students to think about their own thinking, which is an important part of making that thinking better.

✔ **Wait for a response.** What goes on in the silent time between asking a question and getting an answer is thinking. One of the most important findings in educational research is that increased *wait time* — the time between a teacher asking a question and the next time someone speaks — is strongly associated with increased learning. When teachers give their students more time to think about their questions, students learn more. It's true at home, too. Ten or 15 seconds seems like a long time to sit silently when *you* know the answer, but it's not long at all to the person trying to figure out the answer.

✔ **Share a strategy**. After your child explains his thinking, talk about your own. Don't tell him how he needs to do something; just tell him in the spirit of sharing your own ideas. It's like being at the dinner table. If you want your child to share something that happened during the day, you should model it by sharing your own stories from the day. It's the same with thinking. If you want your child to engage with math homework, you can model that behavior by talking about how *you* think about these problems.

Part II
Focusing on Elementary Math: Kindergarten through Fifth Grade

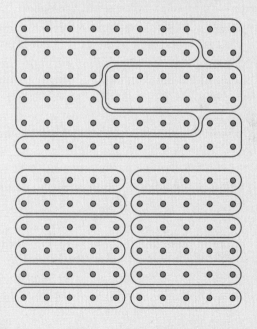

In this part . . .

✔ Grasp the math that your child is studying in school this year so that you can be helpful when he asks you for help.

✔ Master standard algorithms, number lines, and other topics that parents are talking about so that you can spread the word about children's math learning.

✔ See how the math standards develop over the elementary grades so you'll know what to expect your child to learn and when.

✔ Get to know the major topics in elementary math — numbers, operations, algebra, geometry, measurement, and data — so that you have the big picture of each grade's work.

Chapter 5

Beginning with Kindergarten Math

In This Chapter

▶ Comprehending the importance of play in learning math

▶ Seeing the significance of counting

▶ Understanding that kindergarteners put together and take apart both numbers and shapes

*K*indergarten is a critical time in child development. Kindergarteners are still young children who need lots of imaginative and physical play. Yet they also have active and curious minds. A well-run kindergarten classroom should be capable of meeting all these needs. Learning math at this age doesn't need to be about sitting at a desk for long periods of time. Rather, kindergarteners can explore numbers, shapes, and data as natural ways of exploring, representing, and enhancing their play. This chapter gives you an overview of the math children learn as they work and play in kindergarten.

Counting by Ones and Tens

Five-year-olds love to count. They love rhythm and patterns, and they're eager to do what they see the adults in their worlds do. In kindergarten, studying numbers involves all this. As they learn to count to 100 (and beyond), kindergarteners play with lots of number ideas — including the meaning of numbers — and they learn to compare.

Counting is a surprisingly complex skill. Children need to learn the sounds and the patterns of the counting words as well as what those words and patterns mean. These sections describe the difference between saying numbers and knowing how many things there are.

Representing numbers: Counting and cardinality

When you count a collection of things, the last number you say is the total number of things. Saying the number words in the right order and naming one number per object counted is the process of *counting*. Meanwhile, *cardinality* refers to knowing that the last number you mention tells you something about the whole collection.

If you have spent time doing math with small children, you may have witnessed something like the following exchange:

> **Adult:** Can you count these grapes?
>
> **Child:** [carefully pointing to each grape, saying one word for each grape] One, two, three, four, five, six, seven, eight.
>
> **Adult:** So how many grapes are there?
>
> **Child:** Five!

Other children will count again when asked, "How many grapes are there?" Both of these — giving nonsense answers, and understanding the question *how many* as a request to count — are common responses from children who haven't mastered cardinality.

When you count things, you point to each thing and say a word. You also do this when you show someone a set of crayons — red, orange, yellow, and so on. In the case of the crayon colors, each color word describes (or names) an individual crayon. The last color in the crayon collection is just the last color; it doesn't say anything about the whole set. The fact that numbers work differently is something important and challenging for young children to grasp.

They do learn it, though. They learn it by counting, by talking about their counting and about how many things they have or that they need, and then by counting some more. In order to extend the range of numbers they can count, kindergarteners count by ones and they count by tens. They practice counting (both forward and backward), starting at numbers other than one in order to prepare to solve simple addition and subtraction problems in first grade (see Chapter 6 for more on first-grade math).

Comparing numbers: More, less, and equal

Many important questions require comparing numbers to decide which number is greater, or whether the numbers are equal. A child who can count to 20 may still need to think hard to decide whether 8 is more than 5. Questions that require comparing numbers include the following ones:

> I have 7 crackers. You have 5. Who has more?

> Our class has 25 children. We have 17 copies of a favorite book. Do we have enough for everyone, or will we need to share?

Thinking about these everyday kinds of questions focuses children's attention on the important ideas that numbers can be compared and that numbers can help you make good choices and treat everyone fairly.

Kindergarteners refine their knowledge of comparisons. Young children may think that the word *more* has to do with *how much space* the objects take up, so a dinner plate with peas dispersed all over it has *more* than a plate with the same number of peas in a neat pile, even though the number of peas is *equal* in these two situations. Kindergarteners explore and play with this idea.

Finally, *equal* means *is the same as* is an important idea in kindergarten. When children see many problems of this form: $3 + 4 =$, they may develop the idea that the equal sign means *and now write the answer*. This second meaning is a common misconception that often lingers into middle school or even high school math and can be a source of trouble when students study algebra.

You can encourage your child both to understand the idea of equality and to correctly use the equal sign by talking about problems such as $3 + 4 = \_ + 2$. Ask what number fills in the blank. If your child says 7 (which is a common reply), you can say, "Well, $3 + 4$ is 7, yes. But I think we need to put in a number so that 3 and 4 *make the same amount as* some number and 2." Ask how your child can figure our what *would* make the same amount.

Try using beans to help your child think things through. Say, "If I have 3 beans and 4 more, and if you have 2 beans — how many more do you need so we have the same number?" You can point out that $3 + 4$ and $5 + 2$ are different ways to make the same number: 7.

Focusing on Operations and Algebraic Thinking

Kindergarten students don't just study numbers through counting. They also solve problems, and they play with the properties of numbers. Among these properties are that numbers can be combined, and that — correspondingly — any number can be taken apart into smaller bits. They also study the properties of addition and subtraction. These sections examine these properties in greater depth.

Solving problems by counting

Nearly all children enter kindergarten with some ideas about numbers. Although children's home environment and their experiences with preschool programs — together with natural interests — conspire to create variation among children, almost all arrive with an idea that numbers answer certain kinds of questions and that their fingers are awfully good tools for working with numbers.

In kindergarten, fingers and other physical objects such as blocks help children keep track of quantities as they solve problems. They think of addition mainly as *putting together* — "I had 5 apples, then I picked 3 more." They think of subtraction mainly as *taking apart* — "I had 5 apples, then I ate 3."

When children come to kindergarten, they already know about putting groups together and taking them apart. Their knowledge of addition and subtraction comes from these experiences. Understanding of addition and subtraction is built on top of children's knowledge of everyday situations, not the other way around. Children don't need to learn their number facts before they can solve word problems.

Putting numbers together and taking them apart

Some combinations of numbers are common and important. Children need practice noticing and thinking about these combinations. For example, combinations to make 10 are significant. *Mental math* (that is, figuring out computations in your head) and paper-and-pencil algorithms in first and second grades (refer to Chapters 6 and 7) require knowing combinations to make 10. When students see the number 1, they should think about 9 as the other half of the combination to make 10. When they see 4, they should think about 6, and so on.

Being fluent with these facts helps students find their way around the number system. If they learn in kindergarten that 4 and 6 make 10, then adding $14 + 16$ in second grade is easy. You have one 10 from the 14, another 10 from the 16, and then a third 10 from the 4 and the 6, so $14 + 16 = 30$.

You can also use the idea of making tens when you add $7 + 4$. If you know that 7 and 3 make 10, then you may see that 7 and 4 make 1 more than 7 and 3 do. Therefore, $7 + 4 = 11$. Because the number system is based on 10, being able to combine numbers to make tens — and being able to break 10 apart — is a skill that pays dividends for years to come.

But notice what I did in that $7 + 4$ computation: I broke 4 apart by thinking about 4 as *1 more than 3*. Nothing in the statement of the problem $7 + 4 =$ tells me to break apart the 4 (and I could just as easily have broken apart the 7, because $6 + 4$ is another combination to make 10). Breaking numbers apart is a habit of mind that comes from counting, playing, talking, and listening to lots of different ways of thinking about numbers. All are important activities for kindergarteners to engage in on a daily basis.

Noticing Ten: Place Value

Ten is the most important number. Not because having 10 things is better than having 9 things or 11. No, ten is important because it's the foundation of how you say and write numbers; it's the number at the heart of the place value system (see Chapter 7 for extended discussion of the importance of place value).

Because ten is so important, kindergarteners spend a lot of time noticing it and using it. Kindergarteners may represent ten in a ten frame (see Figure 5-1). A *ten frame* consists of two rows of five squares. In the squares, children can place counters — usually of two colors. In addition to helping students notice 10, ten frames can also help them identify and remember combinations making 10, such as 6 + 4 (refer to the previous section "Putting numbers together and taking them apart" for more details).

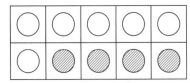

© John Wiley & Sons, Inc.

Figure 5-1: A ten frame showing 6 + 4 = 10.

They may represent 10 with their fingers or toes. They certainly can represent 10 by emphasizing 10, 20, 30, and so on when they count past these numbers, and they count by tens to further emphasize the importance of 10.

Comparing and Classifying: Foundations of Measurement

Kindergarteners notice characteristics of things that can be measured. These characteristics are called *measurable attributes*. Measurable attributes of a person include *height, favorite color, shoe size,* and *bedtime.* Not all measurable attributes can be expressed in numbers — favorite color, for example — so students discuss which can and which can't.

When attributes *can* be measured numerically, students compare with certain types of questions, such as "Who is taller?" or "Who has bigger feet?" These types of questions call on students to compare people to each other according to different attributes.

Students also compare objects according to measurable attributes with these kinds of questions: "Which is the taller tower of blocks?" and "Does a new pencil have longer or shorter lead than a sharpened one?"

Measurement is also about classifying. When you *classify*, you put things into categories. Kindergarteners might classify a handful of toy vehicles by color by putting the blue ones in one pile and the red ones in another. They might classify the same objects in different ways, such as placing the two-wheeled vehicles (bicycles and motorcycles) in one category, the four-wheeled ones (cars and small trucks) in a second category, and the ones with more than four wheels (large trucks and airplanes) in a third category.

Classifying and sorting are natural activities for young children, and the work is closely related to measurement. Being careful about your sorting means naming your categories. And when you name your categories, you're paying attention to a measurable attribute of a number of different objects (for example, *color* and *number of wheels* in the preceding examples).

Finally, kindergarteners compare the number of objects in each category at the end of their sorts. "Are there more two-wheeled vehicles or four-wheeled vehicles?" and "What number of wheels is most common?" are the kinds of questions that kindergarteners can ask after they have finished sorting.

Getting Started with Geometry

Kindergarteners begin geometry by describing shapes while finding them in their world and by playing with shapes by putting them together and cutting them apart. These activities should connect with children's experiences playing and exploring prior to formal schooling, and they should help children develop sophisticated ideas about shapes. It's still kindergarten, though. This should be fun! The following sections closely look at how kindergarteners explore shapes.

Describing shapes

Nearly all children will come to school with ideas about what shapes look like and basic names for shapes. Kindergarteners spend time trying to describe shapes precisely.

A student entering kindergarten may think of squares and rectangles as unrelated shapes. Several reasons may explain this way of thinking. Many shapes books for young children have separate pages for squares and for rectangles. Some of these books even emphasize (incorrectly) that *a rectangle has two short sides and two long sides.* Children may have absorbed these messages. Additionally, as young children identify shapes and the corresponding vocabulary, they may notice that adults tend to use the words *square* and *rectangle* to describe different things.

Kindergarteners learn to identify the differences between squares and rectangles. When studying geometry in kindergarten, they focus on what squares and rectangles have in common. For example, they each have four sides. They can then understand that a square is a special rectangle, not a completely different kind of shape (but don't worry if your kindergartener struggles with this idea and continues to insist that a postage stamp is a square and not a rectangle — this relationship is a challenging one for young minds).

Another important concept that kindergarteners develop by describing shapes precisely is that the orientation of a shape doesn't matter. A square is still a square, even when it's standing on its corner, as Figure 5-2 shows.

Children commonly think of the shape on the left in Figure 5-2 as a *square* and the figure on the right as a *diamond* because they usually see squares presented with a horizontal base, while nearly all four-sided figures presented with a vertex at the bottom of the page are called diamonds. This is true in most shapes books, and in everyday conversation. It's even true in sports. Have you ever wondered why it's called a *baseball diamond* and not a *baseball square*?

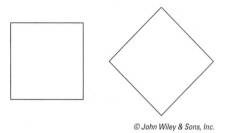

Figure 5-2: Both shapes are squares.

Sit down at a table next to your kindergartener. Have something square shaped handy, such as a sticky note or a postage stamp (don't use a square drawn on a piece of paper because then the paper serves as a reference for the square's orientation). Ask your child what shape the sticky note is. Then turn the sticky note a little bit so that one vertex is closest to you and ask whether it's still a square. Alternatively, you can ask what shape it is now. In another version of this activity, you can put a square on the floor and view it together from a number of different angles. In the first activity, the square changes orientation. In the second one, the square stays put and you change orientation. This difference may matter to your child.

Don't worry if your child responds in ways that aren't mathematically correct (for example, if she says that the sticky note is no longer a square when it stands on its point). This activity is intended to help you listen to your child's mathematical ideas and to help her think about whether the shape's orientation matters. In fact, don't be surprised if she comes back a few days or a week later to tell you something new about tilted squares; children often think about this kind of conversation long after you (or the teacher) have forgotten it.

Playing with shapes

Spatial visualization refers to the ability to picture shapes in your mind and to imagine what they would look like from different perspectives and after being transformed in various ways. Students with strong spatial visualization skills tend to do better in mathematics — in geometry especially, but in mathematics more generally. The important thing to understand about this relationship is that spatial visualization skills can be taught.

In kindergarten, a major predictor of spatial visualization skills is the kind of play children have participated in. Children who have done a lot of building — with blocks or other building toys — tend to have better spatial visualization. Likewise, children who have spent a lot of time navigating complicated environments, such as soccer fields full of children, challenging playground equipment, or paths through the woods, perform better. Children without these kinds of experiences tend to have weaker skills in this area. Studying spatial visualization skills can have a profound effect on both groups —further developing the skills in those with a lot of experience and kick-starting this kind of thinking in children without it.

In kindergarten, students develop spatial visualization skills through active play at recess (throwing and kicking balls, and playing with sand are good for math learning) and through free play in the classroom with blocks and other building toys. They can develop these skills through puzzles and games as well.

Kindergarteners also study putting together (*composing*) and taking apart (*decomposing*) shapes. They notice that two copies of a right triangle make a rectangle (refer to Figure 5-3a) — even if they don't have the language of *right triangle* just yet. They notice that they can cut any polygon into triangles, but the triangles may not be the same size or shape as in Figure 5-3b.

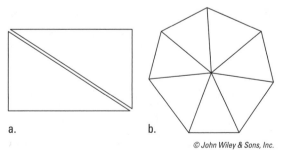

a.   b.

Figure 5-3: Two copies of a right triangle make a rectangle (a), and a polygon can be cut into triangles (b).

Chapter 6

Solving Problems in First-Grade Math

In This Chapter

▶ Understanding keyword strategies and their place (or not) in first-grade math

▶ Exploring problem types for addition and subtraction

▶ Discovering the number ten and its special role in first-grade math instruction

▶ Using geometry and measurement in the first-grade classroom

*I*n first grade, children work on important math ideas that lay a foundation for later grades. They move from the counting of kindergarten to taking numbers apart (*decomposing*) and operating on them (*addition* and *subtraction*). Students study the important number 10 and understand its special role in the number system. They get plenty of opportunities to work with 10 in a variety of ways. First-grade math also involves playing with measurement, exploring shapes, and collecting simple data. This chapter takes a closer look at these areas so you can help your children work through these types of problems.

Digging in to Addition and Subtraction

Although first-grade math is easy for most adults, it's challenging for first graders. They're trying new ideas, thinking hard, and doing much more than memorizing their number facts. In fact, they're starting a long journey to algebraic thinking by studying the operations of addition and subtraction. Their understanding of addition and subtraction can provide a foundation for many years to come.

First graders need to figure out how to solve addition and subtraction problems that involve the following:

✔ A variety of real-world situations, such as putting together, taking apart, and comparing.

✔ Whole numbers less than 20. First graders can handle a lot, but fractions can wait until third grade. (Refer to Chapter 8 for a discussion on fractions.)

An important part of this learning involves encouraging first graders to pay attention to how addition and subtraction are related, rather than thinking about them as different operations. Although this work is supported when students write equations such as $5 + 3 = 8$ and $8 - 3 = 5$, the deepest understanding takes place by thinking about, talking about, and working with these number relationships.

The following sections show you how the approach to word problems in Common Core is different from what you probably remember from elementary school and help you understand the importance of decomposing numbers for learning to add and subtract.

Saying bye-bye to keywords

For many people, the most important shift in the Common Core is from thinking about math as *getting right answers* to thinking about math as *a meaningful collection of really good ideas* (see Chapter 3 for additional information). Those ideas need to lead to right answers, of course. But the *ideas* are the focus of teaching.

In this section, I discuss something that will likely sound familiar from your own elementary school experience — keyword strategies for solving word problems. So-called *keyword strategies* have been a common way of learning to solve word problems. For example, the following adage may have been prevalent in elementary classrooms and textbooks pre–Common Core:

> When you see the following words, you may need to add: altogether, total, in all, more.

With this keyword strategy, students were encouraged to look for the keywords in order to translate problems into mathematical expressions. Keyword strategies are a short-cut to getting right answers without having to think about the math, which isn't okay in a Common Core classroom. Students who are taught keyword strategies learn not to think. They focus on looking for shortcuts instead. Most importantly, they only learn to solve problems that come neatly packaged in math textbooks — the kind of problems on which keyword strategies work. Here are two problems to illustrate why this matters:

> **Problem 1.** Rachel has 4 dollars. Christopher has 3 dollars. How much money do they have altogether?

> **Problem 2.** Rachel and Christopher have 7 dollars altogether. Rachel has 4 dollars. How much money does Christopher have?

You can solve the first problem by using keyword strategies. Apply keywords to the second problem, and you'll get a wrong answer. Even worse, though, is something like the following problem, which is an adaptation of a classic math education research problem.

> **Problem 3.** A ship carries a total of 35 sheep and 7 goats. How old is the ship's captain?

This problem contains a nonsense question, but students frequently attempt to answer it by using the given information about sheep and goats. In this case, 42 would be a popular answer, because it seems reasonable, and results from adding the given numbers (you saw the word *total* in there, right?).

One explanation for why students answer this question is that they have been taught not to make sense of problems, but to look for the keyword, to use the given numbers to find an answer, and then to ask whether the result makes sense. Asking whether the result makes sense is important. But you can't know whether an answer makes sense if you haven't determined that the question makes sense. Keyword strategies skip the step of making sense of the question and of the quantities involved.

These types of strategies may sound like a good idea because they make getting right answers to word problems easier for children. In the reality of classroom learning, though, keyword strategies have only short-term benefits. In the long run, using keywords to solve word problems can do long-term harm, setting children up for later difficulties in math.

Instead of looking for keywords, children need to make sense of problems. Fortunately, that comes naturally for nearly all children. University of Wisconsin researchers have documented the addition and subtraction relationships that children tend to understand before they're taught in school. The Common Core refers to them as "adding to, taking from, putting together, taking apart, and comparing" problem types.

Problems 1 and 2 in this section are variations on the putting-together problem types, with the unknown value in different places. Look for opportunities to notice these different situations in your everyday life and talk about them with your child. Following are examples of each problem type:

- ✔ **Adding to:** I have 5 apples. I pick 3 more. How many do I have now?

- ✔ **Taking from:** I had 9 apples. I ate 4 of them. How many do I have now?

- ✔ **Putting together/taking apart:** I have 3 green apples and 8 red apples. How many do I have altogether?

- ✔ **Comparing:** I have 12 apples. You have 9 apples. How many more apples do you have than I do?

In each problem type, it's also important to play with which value is missing. For example, the following problem uses the same values as the *Adding to* problem above, but switches which one is unknown: "I had some apples. I pick 3 more. Now I have 8 apples. How many did I start with?"

The bad news has been that the familiar keyword strategies of old actually get in the way of kids' learning. The good news is that you can help your children by talking about when and why you add and subtract in your daily life. If you get them to think about the situation, not just hunt for keywords, you can help them be more successful in math class and in using math in their lives outside school.

Decomposing numbers

In first grade, students need to begin to think about numbers more deeply. They need to understand that numbers have properties. Just like a ball can be *orange* and *bumpy* (if it's a basketball, for instance), a number can be *even* and *have one digit* (if it's 8, for example).

One important property of all numbers is that they can be *decomposed*, which just means that you can take a number apart, and you can do it in many ways. You can think of 8 as $4+4$, or as $3+5$, or as $9-1$, and so on.

Being able to decompose numbers is important for two reasons:

- ✔ People who excel in math and science tend to be able to decompose numbers in many ways.

- ✔ Even if you aren't going into math or science careers, your ability to do computations in your head (as you go about your daily life surrounded by numbers) is greatly enhanced by being able to decompose numbers.

The good news here is that this skill can be taught through practice. If you aren't good at it now, you can be in a few weeks. If your child is struggling with it now, he can improve in short order by practicing a little bit every day.

Students who quickly recall memorized facts are impressive, but students who can decompose numbers — even if they take longer to reconstruct a given number fact — often achieve at higher levels.

Taking numbers apart and putting them back together is directly useful in addition and subtraction of larger numbers later on. For example, in second grade, students can think of $97+8$ as $100+5$ if they're used to decomposing 8 (and 100). (Refer to Chapter 7 for more on second-grade math.) These skills also apply in algebra and later mathematics.

You can practice with your child anytime you have a small number of objects and a few spare moments. Put eight things in two piles (maybe you keep three and give him five). Count yours; have your child count his. Then change the number in each pile. In a minute or so, the two of you will have practiced all of the ways to decompose 8 (or 9 or 12 or . . .). You can go one step further by recording your decompositions as "3 and 5" or "3 + 5" and comparing all the expressions you got for decomposing 8 (or 9, or 12, or . . .).

Focusing on the Decimal Number System: Place Value Begins

Learning to count is an amazing achievement of the human mind. What seems simple from an adult perspective is actually the result of many difficult ideas coming together. One of the most difficult of these ideas is the *base-ten-place value system,* which is a fancy term for the usual way of writing numbers.

Here I break down those words. *Base-ten* means that number words and the way you write numbers depend on ten and on multiples of ten. *Place value* means that the 2 in 20 has a different value from the 2 in 32. In 20, the 2 represents 2 tens. In 32, the 2 represents 2 ones (or units). This is because the 2 in each number is in a different place. In 20, the 2 is in the tens place. In 32, the 2 is in the ones place.

Place value may be the most challenging idea in elementary school math. First graders spend a lot of time with the most important number in this system — the number 10 — and recognize its importance in the language and notation of the place value number system. Mastering this basic idea of place value provides a foundation for tackling number facts, relationships, and algorithms in later grades.

The following sections examine in more depth how important 10 is to saying and writing numbers and show how children represent 10 in order to understand its importance. They can help clear up why first graders spend so much time drawing pictures of ten, counting by tens, and so on.

Grasping the significance of 10

In first grade, the number 10 gets the attention it deserves. Students will count by tens and decompose ten (see the earlier section, "Decomposing Numbers" in this chapter).

They also decompose other numbers in relation to ten (such as 12 is 10 + 2; 13 is 10 + 3).

Ten is a special number. The system for naming numbers is based on ten. There is a new number name for each multiple of ten: ten, twenty, thirty, forty, and so on. The numbers in between are named with combinations of these words (for example, *twenty-three*). The names for larger numbers reinforce the importance of ten. For instance, one hundred is a group of ten tens, which makes *one hundred* special, so it has a new name. Similarly one thousand is a group of ten hundreds and is named with a new word.

The system for writing numbers is based on ten, too. There is a units place (or *ones* place), a tens place, a hundreds place (recall that 100 is ten tens), and so on.

Representing tens

In first grade, this important role of ten is highlighted in the classroom through a number of representations and activities, including the following:

- ✔ **Linking cubes:** Students may use linking cubes to show groups of ten. They may grab a couple of handfuls of these cubes, link them together in groups of ten, and leave the remaining cubes loose. This exercise helps to emphasize that (say) 26 has two groups of ten and six leftovers (refer to Figure 6-1a).

- ✔ **Base-ten blocks:** Sometimes called *Dienes blocks*, after the mathematician Zoltan Dienes, or *place-value blocks*, these small cubes (refer to Figure 6-1b) represent ones, sticks made of ten cubes represent tens, flat squares made of ten sticks represent hundreds, and so on. Each power of ten gets a new unit. Students don't assemble the cubes to make the sticks but trade ten cubes for one stick or vice versa. Doing so mimics the carrying and borrowing that are part of the standard addition and subtraction algorithms in the United States. *Carrying* and *borrowing* are also called *regrouping* or *trading*; these words refer to those little 1s you probably write above the numbers as you add or subtract on paper. (Refer to Chapter 10 for more on standard algorithms.)

✔ **Circles and lines:** As a quick way to record the work that they do with cubes, students may draw small circles for the ungrouped units and line segments for the sticks of cubes. (Figure 6-1c shows an example of these lines and cubes.) Unlike the cubes, there isn't a 1:10 relationship in these circles and lines. Instead, the circles and lines refer to the 1:10 relationship that students have expressed with cubes.

✔ **Hundreds chart:** A hundreds chart is organized as 100 squares in ten rows and ten columns. The numbers 1–10 are in the first row, 11–20 in the second row, and so on with 100 in the bottom right corner. Figure 6-2 shows an example of a hundreds chart.

Many interesting ten-related patterns are in the hundreds chart. For example, the ones digit in each column is constant, while the tens digit is constant in each row (except for the last number in each row). As students begin to work on addition and subtraction, they can use a hundreds chart to notice that adding ten is equivalent to moving down one row. Therefore doing so leaves the ones digit constant while increasing the tens digit by 1.

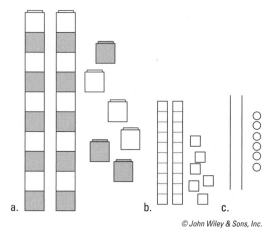

© John Wiley & Sons, Inc.

Figure 6-1: The number 26 is represented with linking cubes (a), base-ten blocks (b), and circles and lines (c).

1	2	3	4	5	6	7	8	9	10
11	12	13	14	15	16	17	18	19	20
21	22	23	24	25	26	27	28	29	30
31	32	33	34	35	36	37	38	39	40
41	42	43	44	45	46	47	48	49	50
51	52	53	54	55	56	57	58	59	60
61	62	63	64	65	66	67	68	69	70
71	72	73	74	75	76	77	78	79	80
81	82	83	84	85	86	87	88	89	90
91	92	93	94	95	96	97	98	99	100

© John Wiley & Sons, Inc.

Figure 6-2: A ten-by-ten hundreds chart.

Working with All Sorts of Measurements

Measurement is an important first-grade concept. For the most part, first graders measure things they touch or experience daily, such as including lengths of familiar objects, time, and money. Collecting data is also part of the first-grade measurement standards. These sections take a closer look at these areas of measurement and offer some helpful ways you can assist your child.

Measuring length

First graders spend a good amount of time studying length. They use blocks, footsteps, pencils, and other physical objects lined up next to the things that they're measuring. They may report to you, "My book is eight blocks wide" or "I am seven shoes tall." First graders are figuring out that the techniques of measurement matter (no gaps between the units, for example) and that different-sized units will result in different measurements.

In the Common Core Standards, this process of laying out physical units and counting is referred to as *iteration*. This word refers to repeating the same unit several times instead of using a ruler that is already marked off in these units. Having lots of experience iterating units is important for understanding how to use a ruler in later grades.

Ask any middle school math teacher this question: Are you happy with all of your students' ability to measure things with a ruler? Then prepare yourself for a deep sigh and an eye roll. When young children get more practice laying out and counting physical units, they're better prepared for ruler use later on. In many elementary classrooms, teachers have expected children to use rulers before they really understand what measurement is all about.

Working with time and money

First-grade students are also working with time and money. You can help with these topics by talking about each of them frequently.

Invite your child to keep track of whether you really do take "just a minute" when he is waiting for you. With a digital clock, he can notice whether the number changes. If you have an analog clock in the house, he can watch the second hand make one trip around. Have your child tell you where the hour hand is or what number is in front of the two dots on a digital clock. Frequently focusing your child's attention just on time is helpful so that he becomes familiar with how time structures our days.

Although not specifically mentioned in the Common Core at first grade, money is commonly taught, and it's age appropriate. By counting small quantities of pennies, nickels, and dimes together, you can reinforce your child's learning of counting words and of skip counting (*skip counting* means counting by twos, threes, or some other number — skipping the numbers in between). (Common Core Standards really start focusing on money in second grade, as I discuss in Chapter 7.)

Measuring the concrete: Data

First graders also collect concrete data. Typically, first graders collect data about themselves. "What kind of pet do you have?" and "Do more people in our class walk to school or take the bus?" are the kinds of questions first graders ask in order to begin thinking about data collection.

After they ask these questions, first graders count and compare the number of answers in each category: "How many dog owners are there?" and "How many more dog owners are there than cat owners?

Before eating a small bag of candy, have your child sort it by color. Lay out the same-colored candies in *bars* (like in a bar graph) and then record on graph paper. Doing this activity with the candies before making the graph on paper can help your child understand that the length of each bar represents the number of each kind of candy in the bag.

Delving into Basic Geometry

First-grade geometry is rather simple, but there is an important twist that you may not notice. Students identify, explore, and work with shapes, which is the simple part. The twist? They work with *defining attributes*, which basically means that they begin to pay attention to what makes a triangle a triangle, not just what a triangle looks like.

When your child was very young, you probably read a number of shapes books with him. A typical shapes book has a page for a triangle with a number of different triangles scattered around the page. These triangles may have been colorful, and they may have been seen in everyday objects (a sandwich, a window, a pizza slice). Almost certainly, these triangles have one side parallel to the bottom of the page.

What children tend to learn from these books is what a triangle *looks like*. A triangle looks like a sandwich cut in half, for example, which is important. The next step is to learn what makes all triangles alike; namely that they have three sides. It's not looking like a sandwich or pointing up to the top of the page that makes a triangle a triangle. A triangle is a triangle because it has three sides and three corners. So expose your child to a variety of triangles in different orientations and talk about whether they are all triangles.

For example, many children will say that shapes A and B in Figure 6-3 are triangles, but that shapes C and D aren't. These children are paying attention only to the *look* of the triangle, not to its properties.

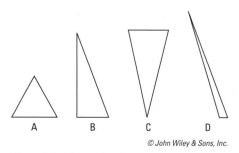

© John Wiley & Sons, Inc.

Figure 6-3: All are triangles.

First graders are also getting better at *spatial visualization,* which is seeing how shapes fit together and imagining what shapes look like from different viewing places. Spatial visualization is an important part of learning about both geometry and numbers. (Number lines and arrays are two examples of using space to represent numbers; see Chapters 8 and 9 for additional information.)

Sit across a table from your child, each of you with a pile of blocks. Build a simple building (maybe three blocks high and one block in front). Have your child copy your building. Pay attention to how he builds his. Where does he put that fourth block? Is it in front from his perspective or from yours? Take turns building, copying, and talking about how your results are alike and different. This exercise is good and fun spatial visualization practice.

First graders go beyond naming shapes in order to understand their properties and to build and imagine with them. Doing so builds a foundation for thinking about categories of shapes in later elementary grades, where students figure out whether a triangle can be both right and obtuse.

Chapter 7

Prioritizing Place Value in Second Grade

· ·

In This Chapter

▶ Understanding how second graders solve addition problems

▶ Seeing the role of equal-sized groups in learning place value and in getting ready for multiplication

▶ Gaining insight into the importance and challenges of place value

▶ Knowing what else second graders study — besides place value

· ·

*I*n second grade, children continue to study geometry and measurement, but the major focus is place value. *Place value* refers to the fact that the *2* in 32 has a different value from the *2* in 23 — and that the only difference between those two 2s is their location in their respective numbers. The 2 in 32 is the farthest digit to the right, whereas the 2 in 23 is second from the right. When you write numbers, location matters. This idea is simple to state, but it leads to complicated results and is surprisingly challenging to learn. Place value is an important and complex idea worthy of the time it takes second graders to learn.

In addition to place value, second graders also continue their work with addition and subtraction. This chapter explains all the different topics that second graders study.

Adding, Subtracting, and Looking for Groups

In second grade, children continue to focus on the operations of addition and subtraction. They develop strategies for solving addition and subtraction problems, including counting, using number facts, writing equations, and drawing diagrams. They also begin to look for same-sized groups in order to start thinking about multiplication in third and fourth grades (see Chapters 8 and 9 for more information). These sections describe students' continued study of operations, including multi-digit addition and subtraction and laying the foundation for multiplication.

Thinking about addition and subtraction

Second graders solve addition and subtraction problems that come mostly in familiar contexts, such as *putting together* and *comparing* (see Chapter 6 for more on these problem types). They may make drawings to show their thinking, and they certainly write equations to show the relationships in the problems they solve.

Learning about addition and subtraction includes understanding what these operations mean. Second graders need to know their single-digit addition facts, and just as importantly, they need to know what situations call for the use of addition and subtraction — by comprehending a problem situation (and not with keywords).

Knowing number facts doesn't mean instant memory recall for all possible combinations. After using certain addition facts (such as those involving 10, or the doubles — $3 + 3$, $4 + 4$, and so on) many times in the process of solving problems, these facts will become familiar to most students and they'll recall them instantly when needed. For other facts (such as $8 + 7$ or $9 +$ almost anything), students usually develop strategies for knowing them quickly. For $8 + 7$, students often think about it as one less than $8 + 8$. This kind of thinking leads to much more powerful math than focusing efforts on memorizing, and it often leads to students committing the facts to memory. Thinking is a better drill than flashcards.

Consider this problem that would be reasonable in second grade:

> Tabitha needs $1.14 to pay for a bag of cheese puffs. In her coin purse, she has a dollar bill and two dimes. Does she have enough money to pay for the cheese puffs? If so, how much change should she get back?

A common way to solve this problem is to notice that the dollar bill covers the $1 part of $1.14. Then a child can notice that the two dimes make 20 cents, which is more than the 14 cents that Tabitha needs.

Now the second grader needs to answer this question: "How much more is 20 than 14?" Some represent it with an equation such as $14 + \_\_ = 20$ or $20 - 14 = \_\_$. An adult may know that these two are equivalent, but most second graders don't immediately identify the similarity.

A child who writes $14 + \_\_ = 20$ is likely to count up from 14, "15, 16, 17 .... " Usually, she keeps track of her counting with her fingers — raising one finger when she says 15, a second when she says 16, and so on until she gets to 20 with six fingers.

By contrast, a child who write $20 - 14 = \_\_$ usually works a lot harder. She usually counts back from 20, "20, 19, 18, 17 ... " until she has counted 14 times and reaches 6. Keeping track of 14 counts is harder, so this strategy is more prone to errors. Other children count back from 20 to 14 instead of counting back 14 times.

In a Common Core classroom, children discuss these various strategies. They talk about which ones are easier, whether each will always work, and about how the strategies are alike or different from each other. An important goal is for a student who counts back 14 times from 20 to notice that she could find the difference more efficiently by counting back from 20 to 14 instead. Learning to do this involves learning more about subtraction as an operation; subtraction tells about *distance* and *comparison*, not just *take away*. Meanwhile, a student who already has multiple strategies should be encouraged to think about using combinations to make 10 (refer to Chapter 6) knowing that 14 (ten and four ones) plus six will make 20 (two tens).

The goal is not for all children to progress in lockstep, which is unrealistic. Instead, the goal is for all children to make their thinking better.

An equation, though, is only one way of representing the situation, "How much more is 20 than 14?" Some students may use counters or make a drawing. Figure 7-1 shows a typical drawing.

ooooo ooooo ooooo ooooo

●●●●● ●●●●● ●●●●

© John Wiley & Sons, Inc.

Figure 7-1: A typical drawing for figuring out how much more 20 is than 14.

Seeing why units matter

When you count groups of things (such as pairs of shoes), you change units. A *unit* is a thing that you count. You can count the number of shoes (one unit) in your closet, or you can count the number of pairs (a different unit) of shoes. Paying careful attention to units is important in setting the stage for multiplication and for understanding place value (refer to the later section "Focusing on Place Value" in this chapter for additional information).

In third and fourth grade, students study multiplication. The foundation for multiplication is equal-sized groups (check out Chapter 8 for more details). To prepare for this work, second graders identify and work with equal-sized groups — even before they name the multiplication and division ideas that follow.

One of the most important equal groupings is a *pair.* Children have lots of experience with things that come in twos. Shoes, eyes, and partners in class are familiar examples of pairs to students. Second grade builds on this familiarity by having students:

- ✔ Count by twos

- ✔ Separate groups of objects into pairs

- ✔ Separate groups of objects into two equal-sized groups

- ✔ Decide whether numbers are odd or even

You can help your second grader notice that some things usually come in groups. Give your child practice counting both groups and individual things. Eggs, bicycle wheels, and grapes are all things that usually come in groups. Ask whether these groups are always (or almost always) the same size. For example, eggs almost always come in 12, bicycle wheels almost always come in pairs, but the number of grapes in a bunch can vary widely.

Whether students are studying groups of two or groups of a different size, one of the more useful ways of showing groups is called an array. In math class, an *array* is a series of things arranged in rows and columns. Figure 7-2 shows an example of an array.

○　○　○　○　○　○

○　○　○　○　○　○

○　○　○　○　○　○

○　○　○　○　○　○

© John Wiley & Sons, Inc.

Figure 7-2: An array of dots.

Arrays are useful because they show two ways of grouping the dots. The array in Figure 7-2 has four groups of six if you consider the rows, or six groups of four if you consider the columns. In third grade, arrays help justify the *commutative property of multiplication* — namely, that the order of the numbers you multiply doesn't matter when finding the total.

I make this distinction because when students begin to study multiplication, they think about situations, such as 4 plates of 6 cookies each. It isn't obvious that 4 plates of 6 cookies is the same total number of cookies as 6 plates with 4 cookies each, and the two situations are quite different otherwise. Knowing that A groups of B is always the same as B groups of A, so $A \times B = B \times A$ is an important achievement. This insight becomes especially useful as the study of multiplication is extended to multi-digit numbers in third and fourth grades, to fractions in fifth grade, and to algebra in sixth grade and beyond.

Many children don't naturally see rows and columns. Given an array, they may haphazardly count the objects around the edge and then in the middle, which often leads to double counting and skipping things. If your second grader doesn't use the rows-and-columns structure of arrays, don't worry, but if she doesn't, then it's a sign that she needs some help noticing this structure.

To help your child notice the rows and columns structure of an array before she knows about multiplication, you can offer lots of examples. After your child counts an array, you can count it in deliberate rows or columns, perhaps saying, "I'm going to count across the top row first …. Now I'll count the next row," and so on. Have her arrange things, such as stickers on a sheet of paper or crackers on a baking sheet for snack in rows and columns. As your child becomes familiar with the structure, you can skip count. If each row has three cookies, you can count by threes: 3, 6, 9, and so on. Point to each row as you say the next number together.

Focusing on Place Value

Place value is the most important idea in second grade. Children learn this skill through exposure and through practice. In far too many textbooks pre–Common Core, the study of place value was limited to naming places, which robbed many children of the opportunity to know place value well and to learn algorithms with meaning. In a Common Core classroom, second graders study place value in ways that allow them to learn it well, which provides a stronger foundation for later arithmetic and algebra learning.

The usual way of writing numbers is a *place value number system*. In other words, a limited set of symbols (called *digits*) builds numbers (0, 1, 2, 3, and so on up to 9) and you can write all numbers using these symbols. Most importantly, the values of these symbols change depending on where they appear in the number. In other words, their location (place) determines their worth (*value*).

As an example, consider these three numbers: 346, 463, and 634. These numbers all have the same digits but have very different values. The *6* in 346 represents *six*. Six ones. But the *6* in 463 represents *sixty*. Six tens. In the number 634, the *6* represents *six hundred*. In each case, it's six *things* — *six units* — but the unit you're counting changes depending on where the *6* is in the number.

As an adult, you're probably so accustomed to this system that this concept seems obvious. But place value is a tremendous achievement of the human mind. People used numbers for thousands of years before inventing place value systems, which means that the rules and behavior of place value number systems aren't obvious at all.

In fact, the development of the number zero was a huge achievement that needed to take place before place value number systems could exist. Without zero, how do you tell *6* from *60*?

Comprehending the place value system involves more than knowing the names of the places. It also involves being able to use relationships among the places.

For example, you can decompose 346 (see Chapter 6 for more on *decomposing* numbers) as $300 + 40 + 6$ or three hundreds, four tens, and six ones. But you can decompose it in several other ways using the hundreds, tens, and ones places. Some examples include

- ✔ 0 hundreds, 34 tens, and 6 ones
- ✔ 3 hundreds, 0 tens, and 46 ones
- ✔ 2 hundreds, 14 tens, and 6 ones

You can think of these in terms of money. There are many different ways to make $346 using $1, $10, and $100 bills.

Each of these alternate ways of thinking about 346 with place value may be useful at one time or another. Especially important is the last one in the list. If you want to subtract $346 - 174$, you might use the standard algorithm for subtraction (see Chapter 10 for more on standard algorithms). If you do use the standard algorithm, you'll end up needing to regroup — or as you may know it from elementary school — *borrowing* or *carrying* in the tens place, which refers to the step where you cross out the 3 in the hundreds place and write a little 1 next to the tens place, as demonstrated here.

$$\begin{array}{r} {}^{2}\mathbf{3}^{1}46 \\ -172 \\ \hline 174 \end{array}$$

When you do that, you're rewriting 346 as $200 + 140 + 6$, which is the same as the third bullet in the previous list. Children who can see a number such as 346 in all of these different ways actually do better with arithmetic, and later with algebra, than children who can only name the places for each digit. These children make fewer errors in their computations, and they're more able to extend their understanding to decimals such as 0.85 and to fractions.

Going Deeper with Measurement

Second graders measure, estimate, and compare lengths using standard units, and they work with time and money. This section describes how second graders deepen their understanding of measurement by considering length, time, and money.

Measuring length

In second grade, students use rulers to measure lengths by connecting and extending their use of nonstandard units such as paper clips and linking cubes in first grade. A ruler is a complicated object to use correctly, so you can expect your second grader to be working on this skill for a few years, gaining accuracy over time.

The basic principle of measuring length is that you compare the length of the object that you're measuring to a standardized length (called a *unit*). If you measure a toy car to be 5 inches long, it's the same length as five 1-inch-long blocks end-to-end. The process of lining up units is called *iterating* (refer to Chapter 6). A ruler is more precise and convenient than lining up units end-to-end because the units are carefully marked and you don't have to worry about leaving gaps between the units.

But this property of rulers also makes them challenging for children to learn to use. Among the rules you need to follow to use a ruler properly are these:

- Start at zero (not one).
- Count the spaces, not the lines.
- The longest tick marks are whole inches, the shorter ones are halves, the next shorter ones are quarters, and so on.

In second grade, students don't have to worry about fractional measurements. They measure using inches, feet, meters, and centimeters. Nonetheless, they'll be curious about (and possibly flummoxed by) all those little tick marks as they use rulers to measure lengths.

Additionally, students in second grade estimate lengths. Being able to make good measurement estimates is a tremendously helpful skill in life. Second graders work on this concept by guessing how many inches (or meters or centimeters or feet) long something is, and then checking their estimates by measuring. An important part of learning to estimate lengths is becoming familiar with *benchmarks*. These are familiar examples of approximate measures, such as that your pinkie nail is about 1 cm wide. This gives a basis for comparison.

Second graders also compare measurements. They compare the lengths of two objects by saying which measurement is greater or less than the other, and by how much. (For example, "My leg is longer than my arm by 3 inches.") They measure the same thing twice — once in inches and once in centimeters — and discuss the relationship between these results. (For example, "My pencil is about 15 centimeters long and about 6 inches long because centimeters are smaller than inches.")

Working with time and money

Students refine their time-telling skills in second grade. They're expected to tell time in five-minute increments on both digital and analog clocks, and to use and understand a.m. for morning and p.m. for afternoon and evening.

With money, they solve problems that involve dollar bills and any of the major coins: pennies, nickels, dimes, and quarters. The following activity gives you a chance to practice counting money with your child while also reinforcing place value.

Grab a handful of dimes and pennies and a few dollar bills. Have your child count out 54 cents (she will likely do either 54 pennies or 5 dimes and 4 pennies). Show her another way to have 54 cents (say, 4 dimes and 14 pennies). Challenge yourselves to find *all* of the different ways to make 54 cents with pennies and dimes. Then do it again with $1.27 or some other value between one and two dollars.

Identifying and Building Shapes

In second grade, students draw shapes based on their properties. They need to imagine a shape that has a particular set of properties. For example, a student may be asked to "draw a six-sided figure with sides that are different lengths." This task is more challenging than "draw a hexagon."

Consistent with the expectation that students use arrays (refer to "Seeing why units matter" earlier in this chapter) to represent groups of objects, students cut rectangles up using lines (the word *partition* describes this process of cutting with lines) into rows and columns of squares. The resulting figure looks like an *array* of squares. This is an important thing to practice leading to the study of area in third grade.

In second grade, students partition circles and rectangles for another reason — to study basic fractions. Second grade introduces halves, thirds, and fourths as words and as parts of circles and rectangles. (Figure 7-3 shows an example.) The important idea in second grade is that the parts need to be equal-sized in order to be called thirds (or halves, or fourths). The parts do *not* need to be the same shape, though, and second graders can have a lot of fun cutting rectangles into equal-sized, but not sam-shaped pieces. The symbols for fractions, such as $\frac{1}{2}$ and $\frac{3}{4}$, may not be introduced in second grade; they're a third grade standard in the Common Core. (I explain them in Chapter 8.)

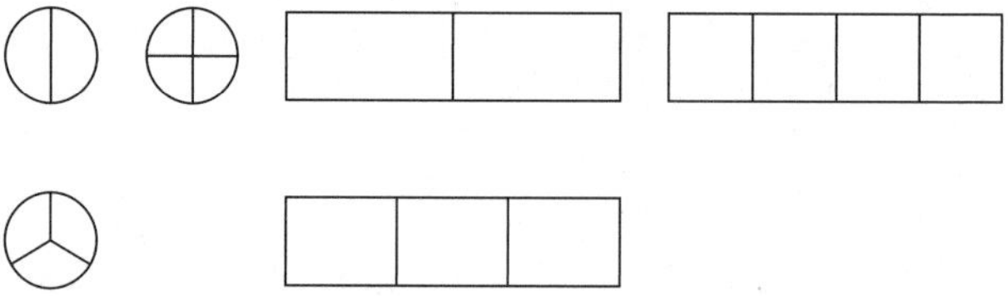

Figure 7-3: Partitioning circles and rectangles to show halves, thirds, and fourths.

Chapter 8

Finding Fractions in Third-Grade Math

In This Chapter

▶ Understanding the meaning of multiplication

▶ Digging into fractions

▶ Looking at measurement and estimation

*T*hird grade is about fractions and multiplication. Students study multiplication and division and the relationship between these operations. They study representations of multiplication, such as arrays and area. They also address fractions, relying on the same idea behind multiplication and division — the idea of same-sized groups. This chapter takes a closer look at these areas and how your third grader stretches her knowledge of multiplication, division, and fractions.

Studying Multiplication and Division

The relationship between multiplication and division is very much like the relationship between addition and subtraction as I discuss in Chapter 6. Whenever you have a multiplication fact, such as $3 \times 2 = 6$, you have two related division facts — $6 \div 3 = 2$ and $6 \div 2 = 3$. For this reason, third graders study multiplication and division together.

The important thing about studying these operations is to develop an understanding of the *meaning* of them at the same time that students are learning their number facts.

The following sections show you the meanings third graders use for multiplication and division, and introduce important properties of these (and other) operations.

Defining multiplication and division

Multiplication is the operation that you use to figure out how many things are in *some number of same-sized groups*. A × B means *A groups of B*. 3 × 5 = 15 because when you have three groups of five things, you have 15 things altogether. In this example, 3 and 5 are *factors* (the two things you multiply together), and 15 is the *product* (the answer to a multiplication problem).

Third graders make same-sized groups, draw pictures of same-sized groups, and use their knowledge of *skip counting* (counting by twos, threes, fives, and so on) in order to find products of whole numbers less than ten.

Students also spend time noticing situations that involve same-sized groups, such as pairs of shoes or dozens of eggs. This helps students understand when multiplication is the right thing to do. Noticing situations is very different from memorizing the keywords that indicate multiplication in a word problem. I discuss keywords in Chapter 6.

Although you can use A × B to find the total number of things in *B* groups of *A* and order doesn't matter when you're multiplying two numbers (check out the next section for more details), the A groups of B convention is useful for studying the structure of multiplication.

Because the two numbers in a multiplication situation represent different things — A is the number of groups of objects, whereas B is the number of objects in each group — the two related division facts have different meanings. These two meanings affect how students tend to think about a division problem and the strategies they may use to solve a division problem.

These two types of division problems are sharing and measuring. A *sharing* problem is one where you know the total and the number of groups, but you need to find the number in each group. A *measuring* problem is the reverse. You know the total and the number in each group but not the number of groups. Here are examples of each problem type:

> **Sharing:** Griffin and his father baked 60 cookies. If five people shared the cookies equally, how many cookies did each person get?

> **Measuring:** Tabitha and her mother baked 60 small cookies. If they put 5 cookies in each box, how many boxes of cookies can they make?

The sharing problem asks about the number of cookies in each of the 5 equal-sized groups. The measuring problem asks about the number of groups of 5 cookies the people can make. The answer to both problems is 12, but the meaning of the 12 is different. What is being counted (the unit) is different. In the sharing problem, it's 12 *cookies* for each person. In the measuring problem, it's 12 *boxes* of cookies.

Before they know their multiplication and division facts fluently, students commonly solve the measuring problem with repeated addition, repeated subtraction, or skip counting. They may think: "5 cookies is one box, then 10 cookies is two boxes, 15 cookies is three boxes," and so on until reaching 60. Solving the sharing problem with these methods is rarer because *five* is the number of groups in a sharing problem, not the number of objects.

By contrast, students are less likely to use repeated addition, repeated subtraction, or skip counting to solve sharing problems. Instead, they're more likely to work in ways that are like dealing cards: "One for you, one for me," and so on. In the example, a student may draw a picture of five stick figures, then draw one cookie next to each stick figure in turn, counting until reaching 60. Each stick figure would then have 12 cookies.

More sophisticated students may use related multiplication facts to solve a sharing problem. In the example, a student may think, "If everyone got 6 cookies, that would be 30 cookies total. There are two 30s in 60, so they can do that twice; everyone gets 12 cookies."

Figure 8-1 shows these two division ideas. Sixty divided by 5 gives either 12 groups or 12 in each group.

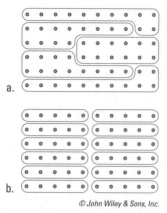

© *John Wiley & Sons, Inc.*

Figure 8-1: Sixty divided by 5 two ways: groups of 12 (a) or 12 groups (b).

Tackling properties

Associative, commutative, and distributive — each word is a mouthful. I tell you what they mean in the following sections, but the words themselves aren't the important thing for third graders. The ideas that they represent are significant. Together, the associative, commutative, and distributive properties make algorithms such as long division work, and they're the foundation on which algebra is built.

Starting simple: The commutative property

Addition and multiplication are *commutative,* which means that $2 + 3 = 3 + 2$, and that $2 \times 3 = 3 \times 2$, and that switching the order of the numbers in either an addition or a multiplication expression gives a true equation no matter what numbers you use. The commutative property of addition says that $a + b = b + a$ for all values of a and b. The commutative property of multiplication says the same thing about multiplication: $a \times b = b \times a$ for all values of a and b.

The commutative property makes it much easier for children to learn their number facts. Instead of needing to memorize all 81 facts on a 9 by 9 multiplication chart separately, they can reduce their load by learning 45 facts and knowing that they can reverse the numbers when they need to.

Not every operation is commutative, though. Addition and multiplication are special this way. Subtraction and division aren't commutative. 5 – 3 isn't the same as 3 – 5, and students need to notice this, too.

Ordering the operations with the associative property

Addition and multiplication are *associative*, which means that $(2+3)+6 = 2+(3+6)$, and that $(2 \times 3) \times 6 = 2 \times (3 \times 6)$. This idea is subtle, because it depends on the parentheses. But you probably use this idea all the time.

Quick, what is 60 × 4?

Did you think "Six times four is 24, then put a zero at the end"? If so, you used the associative property of multiplication. Did you think, "Two 60s are 120, and two 120s are 240"? Again, that's the associative property at work.

You see, 60 is 10 × 6, so 60 × 4 is really $(10 \times 6) \times 4$. That's the same as $10 \times (6 \times 4)$ — the associative property of multiplication. The parentheses indicate that you need to multiply 6 by 4 first, getting 24. And then multiplying by 10 puts the zero on the end.

At the same time, 4 is 2 × 2. So 60 × 4 is really 60 × (2 × 2). That's the same as (60 × 2) × 2. The parentheses indicate that you need to multiply 60 × 2 first, getting 120. Multiplying by 2 again gets the correct product: 240.

The associative property of addition is what younger children use when they compose and decompose numbers (refer to Chapter 6 for more details). For example, $17+8 = 25$, but how do you know that? Some people think to themselves that 8 is $3+5$, so $17+8$ is really $17+(3+5)$. Figure 8-2 shows how you can add three objects to 17 instead of to 5, yet get the same result.

The original sum is the same as $(17+3)+5$, or 25. In this example, the memorized fact $7+3 = 10$ combines with the associative property of addition to make a new fact accessible.

Multiplying with the distributive property

The *distributive property of multiplication over addition* — or the *distributive property* for short — lets you break numbers up into smaller pieces when you multiply, yet get the same answer. Specifically, the distributive property states that $(a+b) \times c = (a \times c) + (b \times c)$.

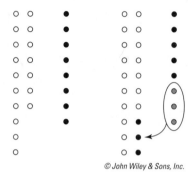

Figure 8-2: $17 + (3 + 5)$ is the same as $(17 + 3) + 5$.

What is 13×9? Can you do it in your head?

Maybe you learned your multiplication facts up to 12. If so, you may remember right away that $12 \times 9 = 108$. So if 12 groups of 9 equal 108, then 13 groups of nine must be 117. If you did this, you used the distributive property — $(12 \times 9) + (1 \times 9)$. Or maybe you don't know 12×9 right away. Another common strategy is to think of 13×9 as ten 9s and then three more 9s — $10 \times 9 = 90$, and $3 \times 9 = 27$, so 13×9 is the sum of these: 117.

An *array* (refer to Chapter 7 for more information) is a useful diagram for representing the distributive property. In Figure 8-3, the whole array is 13×9. The dots to the left of the line show 12×9 and the dots to the right of the line show 1×9. Finding those two products individually and then adding them together gives the same product as the whole array. That's the distributive property.

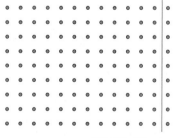

Figure 8-3: An array illustrates the distributive property.

Mastering Addition and Subtraction

In third grade, students work on multi-digit addition and subtraction. The goal is to be able to solve problems such as 345 + 289 and 753 – 168 quickly, accurately, and with meaning.

Students use a variety of strategies to work these problems. Some students may have a favorite way of thinking about subtraction that they use for all problems; other students may change their strategies depending on the problem.

Find this difference: 1,002 – 998. Then find this difference: 132 – 76.

Many people solve these two problems in very different ways. Maybe you did too. A common way to solve 1,002 – 998 is to think, "How far apart are 1,002 and 998? Well, 1,000 is between them. 998 is two less than 1,000 and 1,002 is two more than 1,000. So 1,002 – 998 = 4." If you look at the difference as asking how far apart, then you can probably do this problem in your head.

The second problem, 132 – 76, is harder for most people to do in their heads. Not impossible, but harder (for example, 100 is between them — 76 is 24 less, while 132 is 32 more, so 132 and 76 are 56 units apart).

The second problem is one that many people will get out paper and pencil in order to solve. Figure 8-4 shows the standard algorithm (I discuss standard algorithms in Chapter 10) on the left and a different kind of work on the right.

$$
\begin{array}{r} \overset{2\,1}{1\,3\,2} \\ -\ 7\,6 \\ \hline 5\,6 \end{array}
$$

a.

b.

$$
\begin{array}{c}
132 \ - \ 76 \\
100 \ - \ 44 \\
60 \ - \ 4
\end{array}
\quad
\begin{array}{l}
-32 \\
-40
\end{array}
$$

-32 -40

$\left(\,56\,\right)$

© John Wiley & Sons, Inc.

Figure 8-4: Two solution methods for 132 – 76.

The thinking behind the work on the right is this: You want to take 76 things from a collection of 132 things, so stick to these steps:

1. **Take away 32.**

   Doing so gets you to a nice, round 100.

2. **Take away another 44 things to total 76 things taken away.**

   That means the first line (132 − 76) is equal to the second line (100 − 44).

3. **Take the 44 away in two pieces.**

   First 40 and then 4 more.

To be clear, the work on the right is not the Common Core way of solving this subtraction problem. It is a way of keeping track of a useful thought process — one that involves decomposing numbers (decomposing basically means taking apart without changing the value — see Chapter 6 for details about decomposing numbers).

The use of 100 in this computation is an example of using place value. One hundred is a special and important number in the number system; it's the smallest three-digit number, for example, and it's made of ten tens (and ten is the basis on which the whole number system is built).

You *could* do the first problem (1002 − 998) using the standard algorithm, and you *could* do it using the idea of taking away in several steps, as on the right in Figure 8-4. However, these methods are a hassle because neither one is easy to do in your head, and there are many opportunities for errors.

Insisting that students have or use only one strategy for solving multi-digit subtraction problems pretty much guarantees an increased number of computational errors. Put more simply, kids get things wrong more often when they only have one strategy. Refer to Chapter 10 for more on standard algorithms and their advantages and disadvantages.

Exploring Fractions

In third grade, students study fractions for the first time. Most students know something about fractions from their everyday lives. They know about splitting things in half and have encountered *quarters* in the contexts of both money and time. But they haven't studied fractions as mathematical objects.

The emphasis is on *unit fractions,* which are fractions with a 1 in the *numerator* (the top part of the fraction), such as $\frac{1}{2}$, $\frac{1}{4}$, $\frac{1}{5}$, and so on. Much of the third grade work with fractions occurs on the number line as students cut up intervals and label points on the number line to think about fractions as both locations and lengths on the number line. These sections show you how third graders use number lines to better understand fractions.

Partitioning Cutting up things

Students partition things in third grade, and they name the resulting pieces using fractions. *Partitioning* in this context means cuttings things into equal-sized pieces. Children have lots of practice thinking about how to share something equally. When you cut a large cookie into five equal pieces so that you and your four friends each get the same amount, you make fifths. You call the size of each piece one-fifth, and you can write $\frac{1}{5}$. The number language, the symbols, and thinking about number lines are all new for third graders. Partitioning into equal-sized pieces is familiar to them.

It's important that the pieces are equal-sized, but they don't need to be the same shape. Each of the pieces in the square on the left in Figure 8-5 is one-fourth of the whole square, but that isn't true of the pieces of the square on the right.

The process of partitioning something into b same-sized pieces and then taking one of them gives you a fraction of the form $\frac{1}{b}$, which is called a *unit fraction.* A *unit* in math is whatever you call *one.* This business of units is important. Misunderstandings of units result in all kinds of mathematical errors, so remember this basic principle: A *unit* is the thing that you count.

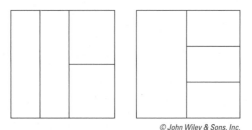

Figure 8-5: The square on the left shows fourths; the square on the right doesn't.

I once observed the following conversation between my two children after the younger one had made brownies with their mother.

> **Older child:** How many brownies did you make?
>
> **Younger child:** One big one. Mommy cut it up.

The older child was thinking that a brownie is the thing that you get when you cut up the pan. The younger child was thinking that a brownie is what you have *before* you cut it up. Younger child may have been thinking that brownies work like cake — you make one cake and then cut it into slices. There is no good reason that in American, English-speaking households, a *cake* is a unit that you partition into slices, while a *brownie* is a unit that comes from partitioning something larger.

Similar miscommunications about units are much less funny when there are consequences involved. When children first learn to count, *one* is the unit. When they count to five, they have counted five *ones*. Later, they learn to skip count. For example, they count by twos: *two, four, six, eight*, and so on. If you keep track on your fingers of the number of times you have counted by twos, your fingers are counting a different unit. Count to ten by twos and you'll have counted five times — that's five twos. Because the unit is the thing you count, this is an example of two as a unit.

When children study place value (check out Chapter 7 for more on place value), they notice that 50 means *five tens*. You can count the tens in 50. Because the unit is the thing that you count, ten is a unit with a special role in the number system. Ten tens make one hundred, so one hundred is a unit. Place value depends on changing units — from ones to tens to hundreds and so on.

Now when students study fractions, they work with unit fractions. The unit fraction is a fraction with a 1 in the numerator, but the *meaning* of a unit fraction is that you can use them to count. Three-fourths means three units of one-fourth. You can count by fourths, just like you counted in kindergarten and first grade: one-fourth, two-fourths, three-fourths, four-fourths, five-fourths, and so on.

This idea is emphasized in third grade. Fractions are built from an original unit by partitioning into equal-sized pieces and then collecting some of these pieces. If you partition into b equal-sized pieces, then collect a of them, it gives you the fraction $\frac{a}{b}$.

Putting fractions on number lines

You may remember a number line running across the front wall of an elementary classroom when you were in school, with a point for zero, a point to the right of that marked 1, then 2 and so on, with an arrow on the end to suggest that the line and the numbers continue forever. Third graders use the number line to build a foundation for fractions.

The two most important numbers on a number line are 0 and 1. After you mark them on the number line, every other number has a defined location. Correspondingly, when third graders use the number line to study fractions, they mark 0 and 1 and then partition the interval between these numbers into same-sized pieces (refer to the earlier section "Partitioning means cutting up things" for more information).

For example, the number line in Figure 8-6 shows the interval between 0 and 1 partitioned into fifths. The mark closest to 0 is labeled one-fifth, which is a unit fraction. The remaining marks are labeled as multiples of one-fifth. Students think about three-fifths as being the third mark to the right of 0 when the interval is cut into fifths.

Figure 8-6: Fifths on the number line.

Third graders basically show numbers on a number line in two ways:

✓ They show a fraction as a location on the number line as Figure 8-6 demonstrates.

✓ They also think of a fraction as a *distance* on the number line. One-fifth isn't just the tick mark labeled $\frac{1}{5}$; it's also the distance between adjacent pairs of tick marks on this number line.

Estimating and Measuring Precisely: Having It Both Ways

In third grade, students measure time more precisely, and they also study area. They also develop their estimation skills. These two ideas — estimation and precision — may seem sort of contradictory. If you need a precise measurement, you shouldn't just estimate it. If you need an estimate, you shouldn't waste your time and effort being precise.

In fact, these two skills are complementary. When you measure, you make mistakes (you know the carpenter's adage: *Measure twice, cut once*). Estimation is helpful for making sure your measurement and computations make sense. But you can't develop the kind of intuition for measurement that estimation relies on without having made a large number of good measurements.

These sections examine these two skills as third graders further develop their measurement skills.

More than just rounding

People who are good at estimation have an ability to answer the question "About how much. . . ?" with confidence. *Estimation* basically is making a rough calculation or an informed and reasonable guess. Sometimes, estimation is presented as being about rounding numbers to the nearest 10 or 100 before computing. There is certainly a place for this, but the heart of estimation is based on a good sense of quantity and measurement.

Answer these estimation questions:

> ✔ About how many hours do you sleep in a week?
>
> ✔ About how much does a gallon of milk weigh?
>
> ✔ About how much does it cost to serve lunch to every student at your child's school on a typical day?

Although round numbers make sense as estimates, none of these questions requires rounding. Instead, answering them requires you to bring in additional information — something you may know, such as the number of fluid ounces in a gallon, or your experience, such as carrying home groceries.

Counting squares to find area

Area is the number of unit squares that fit inside a figure without gaps or overlaps. A unit square may be any size — a square inch, a square centimeter, even a square mile. These squares may be cut into fractional pieces, and ultimately students will be able to think about area without imagining counting squares. But at heart, area is about covering a figure with unit squares.

At first, third graders count these squares to find areas. But soon they develop a formula for the area of a rectangle. Students learn that the area of a rectangle can be found by multiplying the length and the width (both measured in the same units), which the formula $A = l \times w$ summarizes.

The reason for the formula is important. *Why* can you multiply the length by the width of a rectangle to find the area? Why don't you *add* the length and width, for example? Third graders have a reason for this. If you cover a rectangle with unit squares, you can do this with rows and columns. The number of squares in one row is the same as the length of the rectangle. The number of rows is the same as the width.

You can partition a rectangle into an array of unit squares. Because multiplication finds the total in a certain number of same-sized groups, multiplying the number of squares in a row (the length) by the number of rows (the width) gives the area.

A consistent finding of elementary and middle school achievement data is that students often confuse area and perimeter. (*Perimeter* is the distance around a polygon; you can find perimeter by adding all of its side lengths together.) When asked for perimeter, students often find the area and vice versa. When students develop the ideas behind the formula for area — the meaning of multiplication (refer to the earlier section "Defining multiplication and division" in this chapter for more information) and the structure of arrays (flip to Chapter 7) — they should be less likely to confuse area with perimeter. Third graders also study perimeter of rectangles.

Representing data with graphs

Data analysis in third grade is uncomplicated. Students make bar graphs to represent several different categories. Whereas in earlier grades (see Chapter 6, for example) students worked with two or three categories at a time, in third grade their data sets are richer for having more categories. *Kinds of pets*, *favorite foods*, *dream jobs*, and so on are examples of data sets that can have multiple categories for students to compare.

Categorizing Shapes

A square is a special rectangle. Third grade is the place where students learn this concept and other related facts. Children often come to school being able to identify squares and rectangles (and many other shapes) based on appearances, but they usually lack ways of talking about the relationships among the shapes they can identify.

In third grade, students analyze the definitions of shapes and the relationships that result from these definitions. A rectangle, for example, may be defined as a *quadrilateral* (a four-sided figure) with all right angles. A square has these properties and an additional one — all sides are the same length, which means that a square is a special rectangle, not a completely different category of object.

Quadrilaterals have just enough complexity to be interesting to study, so they form the main territory for exploring the ideas of defining and categorizing shapes in third grade.

Chapter 9

Mastering Multiplication in Fourth-Grade Math

In This Chapter

▶ Continuing to work with the four operations of arithmetic

▶ Finding equivalent fractions

▶ Converting units of measurement

▶ Studying angles, lines, and their properties

*T*his chapter explains that fourth grade marks a transition point from the world of whole numbers to the world of fractions. To be sure, students continue to study addition, subtraction, multiplication, and division of whole numbers. But they begin in earnest the study of fractions, including some of the most challenging topics — equivalence and comparison of fractions. Students use fractions to describe and understand other things in fourth grade, especially angles, where the unit of measurement (a *degree*) is defined as a fraction of a turn.

Focusing on Multiplication: Factors and Multiples

In fourth grade, students begin to study the multiplicative structure of numbers. In third grade (refer to Chapter 8), students learn their multiplication facts and solve problems with multiplication. In fourth grade, they study how numbers are built from multiplication relationships.

The important relationship here is between factors and multiples. A number's *factors* are all the whole numbers by which you can divide it with no remainder. For example, the factors of 12 are 1, 2, 3, 4, 6, and 12. Factors come in pairs, so the *factor pairs* of 12 are 1 and 12, 2 and 6, and 3 and 4. Sometimes thinking only about the factors that are less than the number in question is useful. They're called *proper factors*. The proper factors of 12 are 1, 2, 3, 4, and 6.

Meanwhile, the *multiples* of a number are all the numbers that *it* divides with no remainder. The multiples of 12 are 12, 24, 36, 48, and so on. Multiplying 12 by any whole number gives a multiple of 12 —$12 \times 1 = 12$, $12 \times 2 = 24$, and so on.

Every number has a limited number of factors and infinitely many multiples.

The key building blocks of numbers are prime numbers. A *prime number* is a number that has exactly two factors: 1 and itself. For example, 7 is a prime number. The only factors of 7 are 1 and 7. (You can divide 7 by 2, but the answer isn't a whole number.)

Every whole number (besides 1) is either a prime number or it's built out of prime numbers. If a whole number greater than 1 isn't prime, it's *composite*. Numbers such as 6, 9, 10, 12, and 100 are examples of composite numbers. For example, 6 has 2 and 3 as a pair of factors — $2 \times 3 = 6$ — so 6 has more than just 1 and 6 as factors.

In fourth grade, students study the relationship between factors and multiples — namely that a number is a factor of its multiples and is a multiple of its factors. For example, the factors of 6 are 1, 2, 3, and 6. If you list out the multiples of any of these factors, 6 is on that list. The multiples of 2 are 2, 4, 6, 8, and so on. This intimate relationship between factors and multiples often leads to students confusing the terms factor and multiple. But getting the words right every time doesn't matter nearly as much as knowing that numbers can be decomposed with multiplication by using factors.

Calculating with Place Value

Fourth graders continue to study place value. In earlier grades (see, for example, Chapter 5), students studied the structure of the number system, made groups of ten, and used some of the basic properties of place value.

But in fourth grade, students really start to use place value to do things. As the following sections explain, they decompose numbers with place value, and they use algorithms that depend on place value to add and subtract.

Decomposing numbers

Decomposing numbers by place value (check out Chapter 7 for details) and doing so in multiple ways is important for being able to compute efficiently and accurately and for the later study of algebra. In fourth grade, students parlay this place value knowledge into multi-digit computational fluency.

How many tens are in 468?

If your answer is *six*, then you have the right answer to a different question: "What digit is in the tens place?" In fact, many more than six tens are in 468. There are 46 tens in 468 — almost 47 tens! Seeing 468 as 46 tens and 8 ones is one of several useful ways to use place value to analyze this number and others like it. A real-world example of this is money. One way to count out $468 is with 46 $10 bills and eight $1 bills.

Striving for fluency: The case of addition and subtraction

Computational fluency refers to the ability to calculate accurately and efficiently. A student who is fluent with multi-digit addition and subtraction can do many computations mentally and can use paper and pencil to solve problems that she can't do in her head. She may need to think her way through a computation, but she makes few errors and doesn't get too absorbed in computing that she forgets what question she was trying to answer in the first place.

Fluency, then, isn't something ill defined, such as *understanding* is. But it's also not about timed tests or using a particular algorithm.

The standard algorithm for addition is something of a strange one (refer to Chapter 10 for more info). If you were to add two three-digit numbers in your head, such as 245 and 762, you would likely start with the hundreds place. You might think that $200 + 700$ is 900, so your final sum is going to be more than 900. When you add mentally, you usually start on the left, with the largest place value.

The standard algorithm forces you to do the opposite — to start from the right. Doing so is fine, and the standard algorithm works perfectly, although most people don't think about numbers this way. In the standard algorithm, you would start with the smallest place value — here that's the ones place (refer to the following equation). After recording the sum in the units place, you move to the tens place, and then to the hundreds place (and so on, for numbers with a larger number of places).

$$\begin{array}{r} 245 \\ +762 \\ \hline 7 \end{array}$$

The standard algorithm does have alternatives. Your child may learn one or more of them at school. Rest assured that she also learns the standard algorithm (the standard algorithms for addition and subtraction are required at fourth grade in the Common Core).

Forming Fractions

In fourth grade, students get to the heart of studying fractions. They work with unit fractions, which formed the introduction to fractions in third grade (refer to Chapter 8 for more details), and they study equivalence and comparison of fractions. These sections walk you through the fraction talk in fourth-grade math.

Using unit fractions

A *unit fraction* is a fraction that has a 1 in the numerator. (The *numerator* is the top number in a fraction, and the *denominator* is the bottom number.) Example of unit fractions include $\frac{1}{3}$, $\frac{1}{6}$, and $\frac{1}{10}$. In third grade, students used unit fractions as the building blocks for all other fractions, which means that $\frac{3}{4}$ consists of three things — each is worth one-fourth. Three-fourths is the same as one-fourth + one-fourth + one-fourth, much like three dogs is the same as one dog + one dog + one dog.

In fourth grade, unit fractions are the way students understand addition, subtraction, and multiplication of fractions. Addition and subtraction require using the same units. You can't add apples to oranges unless you think of them both as *pieces of fruit.* You can't add hours to miles. Apples, oranges, pieces of fruit, hours, and miles are all examples of units. A unit is the thing that you count, and you can add numbers when they're in the same units. This is another important meaning for unit fraction.

You can think of a unit fraction as a unit, so $\frac{3}{4}$ is three $\frac{1}{4}$s. Then you can add $\frac{3}{4} + \frac{2}{4}$ by knowing that $3 + 2 = 5$, and that the unit is fourths, so $\frac{3}{4} + \frac{2}{4} = \frac{5}{4}$. Rewriting this equation as 3 fourths + 2 fourths = 5 fourths makes it easier to see. In any case, you can't add $\frac{3}{4} + \frac{1}{2}$ directly because the units are different.

Similarly, students use unit fractions to multiply fractions by whole numbers in fourth grade. Students solve a multiplication problem such as $5 \times \frac{3}{4}$ by thinking of $\frac{3}{4}$ as three things. Then $5 \times \frac{3}{4}$ is five groups of three things or 15 things. The things are fourths, so $5 \times \frac{3}{4} = \frac{15}{4}$.

Going beyond reducing: Equivalent fractions

Equivalent fractions make the world go round. Okay, maybe that's not entirely true, but being able to identify and generate equivalent fractions is a tremendously important skill for arithmetic and algebra alike. Fourth graders use pictures and reasoning to write equivalent fractions. Equivalent fractions are any two fractions that represent the same quantity, such as $\frac{2}{3}$ and $\frac{4}{6}$.

You may remember spending lots of time *reducing* fractions, and even getting test questions wrong because you didn't reduce your fractions, but there is absolutely no reason to insist on reduced fractions. A simpler version of a fraction is helpful sometimes. For example, you may not know exactly how much $\frac{314}{471}$ of an hour is, but you can probably work very easily with its simpler form, $\frac{2}{3}$. In that sense, $\frac{2}{3}$ may be a more useful form of the fraction than $\frac{314}{471}$, but it isn't a more correct form. These two fractions mean the same thing.

Overemphasizing reducing fractions — which is the common term for writing an equivalent fraction using the smallest possible whole numbers for numerator and denominator — has set many students up with misconceptions and needless fear of fractions. The term *reduce* is a problem, for example, because it suggests that the reduced fraction is smaller than the original one, which isn't true. Similarly, telling students that they're wrong when they write $\frac{2}{4}$ instead of $\frac{1}{2}$ leads some students to think that fractions are transformed by the process of reducing — that $\frac{2}{4}$ and $\frac{1}{2}$ are somehow different from each other. In fact, they're the same; they're equivalent. In many classrooms, the term *simplify* replaces *reduce* to avoid these problems. *Simplify* suggests that the fraction is in a *simpler form* — using smaller whole numbers in the numerator and denominator—but that it's the same fraction.

Students may generate equivalent fractions by drawing pictures. If you start with the fraction $\frac{6}{9}$ (shown on the left-hand side of Figure 9-1), you can show that $\frac{12}{18}$ covers the same amount of area as $\frac{6}{9}$ by cutting each of the ninths into two same-sized pieces (in the middle of Figure 9-1). You have twice as many pieces in the whole (18 instead of 9), and you have twice as many pieces shaded (12 instead of 6).

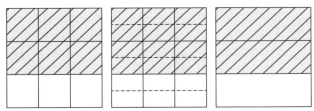

Figure 9-1: Equivalent fractions using squares.

Conversely, you can glue ninths together in sets of three to make thirds. When you do, you divide the total number of pieces by three, and you also divide the number of shaded pieces by three, so $\frac{6}{9} = \frac{2}{3}$, as the right-hand side of Figure 9-2 shows. Your child's teacher or textbook may refer to these pictures as *area models*, because you pay attention to the fraction of the area of the square (or rectangle, or circle, or whatever) that is shaded.

Figure 9-2 shows this same process using a number line. Instead of shading, you keep track of the number of same-sized pieces between the fraction and zero. The top number line shows $\frac{6}{9}$. The middle number line shows the result of cutting each ninth in two equal pieces to get $\frac{12}{18}$. The bottom number line shows the result of grouping the ninths to get $\frac{2}{3}$.

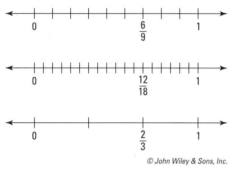

© John Wiley & Sons, Inc.

Figure 9-2: Equivalent fractions using number lines.

After working with equivalent fractions for a while, students develop a rule for writing equivalent fractions. They may express this rule using symbols such as these: $\frac{a}{b} = \frac{a \times n}{b \times n}$ and $\frac{a}{b} = \frac{a \div n}{b \div n}$. In the example of $\frac{6}{9}$, the symbols look like this: $\frac{6}{9} = \frac{6 \times 2}{9 \times 2}$ and $\frac{6}{9} = \frac{6 \div 3}{9 \div 3}$, but this notation isn't required in the standards.

Comparing fractions

Understanding a fraction as a number is important in fourth grade. You use two numbers to write a fraction, but the fraction itself is a number. It has a location on the number line, for example.

If fractions are numbers, then you can compare them. Some numbers are larger than others; children learn this early on. So asking which of two fractions is larger and putting two sets of fractions in order are sensible tasks.

Comparing denominators

The standard way to compare two fractions is to find equivalent fractions that have the same denominator. To compare $\frac{1}{2}$ and $\frac{1}{3}$, you would rewrite them both as sixths: $\frac{3}{6}$ and $\frac{2}{6}$.

Because these two fractions are built of sixths, the fraction with the larger numerator is the larger fraction, so $\frac{3}{6} > \frac{2}{6}$, and therefore $\frac{1}{2} > \frac{1}{3}$, as Figure 9-3 demonstrates.

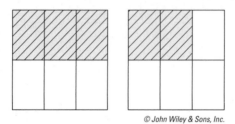

© John Wiley & Sons, Inc.

Figure 9-3: Comparing fractions using common denominators

Comparing numerators

Your child may think about this comparison differently, so here I show you one more way children often compare fractions — a way that likely comes up in a Common Core classroom.

You can compare fractions using common *numerators*. In this way of thinking, you don't need to change the forms of $\frac{1}{2}$ and $\frac{1}{3}$ in order to compare them. With each of these fractions, you have one piece. The question to think about is "How big are the pieces?" If you partition something into two equal pieces, you have bigger pieces than if you partition the same thing into three equal pieces. Partitioning into two equal pieces means each piece is $\frac{1}{2}$ of the original whole. Partitioning into three equal pieces means each piece is $\frac{1}{3}$ of the original whole. Therefore $\frac{1}{2} > \frac{1}{3}$.

A more complicated example could be comparing $\frac{1}{3}$ and $\frac{2}{7}$ using common numerators. You can rewrite $\frac{1}{3}$ as $\frac{2}{6}$ so that it has a 2 in the numerator just like $\frac{2}{7}$ does. Sixths are bigger than sevenths, so $\frac{2}{6} > \frac{2}{7}$, and so $\frac{1}{3} > \frac{2}{7}$. Figure 9-4 shows this problem on the number lines.

Using benchmarks

Another common strategy you may see on homework assignments or that your child may use spontaneously is *benchmarks*. In math, a *benchmark* is a familiar number that you find it easy to compare other numbers to. For many fourth

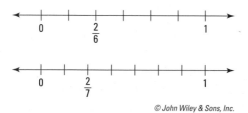

© John Wiley & Sons, Inc.

Figure 9-4: Comparing fractions using common numerators.

graders, $\frac{1}{2}$ is a good benchmark to use. By using this benchmark, students can know that $\frac{2}{3} > \frac{3}{7}$ because $\frac{2}{3} > \frac{1}{2}$ and $\frac{1}{2} > \frac{3}{7}$. In this way, two fractions that are difficult to compare directly can be compared by using a benchmark.

Eyeing Units and Angles

In fourth-grade measurement, students focus on standard and metric units, and they study angle measurement. In geometry, they use angle and side measurements to classify shapes; they consider *right* triangles and *isosceles* triangles, for example. These sections take a closer look at these concepts.

Mastering units of measure

American students are put in a strange place when it comes to measurement — different from nearly every other country in the world. They have to learn two systems of measurement and study the relationships within each of these systems:

- ✔ The standard system of feet, inches, and so on
- ✔ The metric system of meters and centimeters

Studying *relationships within a system of measurement* means that students learn that there are 12 inches in a foot and 3 feet in a yard, and that there are 100 centimeters in a meter and 1,000 meters in kilometer. They don't need to learn to convert *between* systems of measurement — feet to meters, say. (They learn that concept in sixth grade, as a consequence of studying ratios.)

In fourth grade, the emphasis is on multiplying to make conversions from larger units to smaller ones. Students work with questions such as "How many inches are in 5 feet?," "How many centimeters are in 3.5 meters?," and "How many ounces are in 4 pounds?"

Students work with the following units in fourth grade (and possibly additional ones that aren't on this list):

- ✔ Miles, yards, feet, and inches
- ✔ Kilometers, meters, and centimeters
- ✔ Kilograms and grams
- ✔ Pounds and ounces
- ✔ Liters and milliliters
- ✔ Hours, minutes, and seconds

Make a habit of noticing measurements around the house. You don't need to be an expert. The more exposure your child has to the two systems of measurement, the better prepared she will be to build on her experiences in school.

To every angle turn, turn, turn

In geometry, two rays with a common vertex form an angle, as Figure 9-5 shows. A *ray* is half of a line. A ray has one endpoint (the *vertex*) and extends infinitely in one direction.

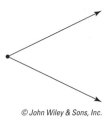

© John Wiley & Sons, Inc.

Figure 9-5: An angle is two rays with a common vertex.

This mathematical definition is usually not that useful for fourth graders (nor even for many 16-year-old geometry students). Fortunately, fourth graders have many experiences with angles that they can build on in the classroom. In particular, fourth graders know about turns. They have spun on the playground, they have tried to turn all the way around while jumping in the air, they have sat on swivel stools, they know that owls can turn their heads much farther than people can, and so on.

When you measure an angle, you measure the size of the turn between the two rays.

Gather a few neighborhood children and do some turns through various angles. Can you turn $\frac{1}{4}$ of a full rotation to the right? Can you turn $\frac{3}{4}$ of a full rotation to the left? If you can do each of these from the same starting direction, the *result* is the same. But the angles are different. The first one is a 90° angle; the second one is a 270° angle.

The $\frac{1}{4}$ turn to the right technically has a measure of *negative* 90°. In the fourth grade, this distinction isn't important. (In high school geometry and in calculus, it becomes useful to be able to describe both the size of the turn and its direction.) Mathematicians have agreed that turns to the left (counterclockwise) are positive, whereas turns to the right (clockwise) are negative. Unfortunately, it's the opposite in navigation, where the standard positive angle measurements are clockwise.

Fourth graders pay attention to fractions of a full rotation — halves, quarters, and so on. They also develop angle measurement in degrees, where a degree is $\frac{1}{360}$ of a full rotation. They use protractors to measure angles to the nearest whole degree.

Addressing Lines and Angles

In fourth grade, students' view of geometry expands beyond shapes. In fourth grade, they think about lines, angles, and the relationships that they can have with each other. This activity generates quite a bit more vocabulary.

For example, students study parallel lines, which are two lines in a plane that don't intersect. Students will think about parallel lines as *never meeting*, or maybe as *headed in exactly the same direction*.

Students begin to make distinctions between *lines* (which extend infinitely in either direction), *rays* (which extend infinitely in only one direction), and *line segments* (which have finite length and two endpoints). Students should investigate whether the definition of parallel lines applies to line segments (see Figure 9-6). In fact, it doesn't — line segments may not meet even if they aren't parallel — so a slightly different definition is required. You can say that two line segments are parallel if they *would never meet*, even if extended infinitely.

© *John Wiley & Sons, Inc.*

Figure 9-6: These line segments don't intersect, but they aren't parallel.

Students use line and angle relationships such as parallel and *perpendicular* lines (*parallel* lines are in the same plane as each other, run in the same direction, and never meet, even if extended infinitely; *perpendicular* lines meet at right angles), and right and acute angles (a *right* angle has a measure of 90°; an *acute* angle has a measure greater than 0°, but less than 90°) to classify and relate shapes to each other.

Finally, students recognize *lines of symmetry* in shapes. If a shape is drawn on a piece of paper, folding the paper along the *line of symmetry* causes two halves of the shape to match up with each other perfectly. A square has four lines of symmetry. A non-square rectangle has only two lines of symmetry (refer to Figure 9-7 for an example). Although you can connect the corners of a rectangle and get two same-sized halves, folding along this line won't make the two halves match up.

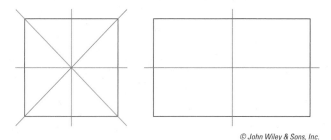

© *John Wiley & Sons, Inc.*

Figure 9-7: A square has four lines of symmetry; a non-square rectangle has two.

Chapter 10

Anticipating Algebra in Fifth-Grade Math

* *

In This Chapter

▶ Shifting from arithmetic to algebra

▶ Moving to the right of the decimal point

▶ Understanding the role of standard algorithms

▶ Looking at adding, subtracting, and multiplying fractions

* *

*F*ifth grade marks the beginning of the transition from arithmetic to algebra. Although much of what has come in previous grades has laid a foundation for this change, the focus prior to fifth grade has been on arithmetic. In fifth grade, students start wrapping up their arithmetic work and getting a first glimpse of algebra.

Part of this transition involves using algebraic symbols (*x* and *y*, for example), of course, but it also involves looking for the structure of a computation. Another part of the transition involves tying up some loose ends of arithmetic, so fifth graders study decimals, and they multiply and they begin to divide fractions. In geometry, fifth graders move to three-dimensional measurement, and they plot points in the coordinate plane (which is also setting them up for studying algebra in later grades). This chapter explains these concepts in more detail.

Expressing Relationships: Early Algebra Ideas

Early algebra ideas include using variables and noticing the structure of computations. (*Variables* are values that are unknown, or that can change, whereas noticing the *structure of computations* means paying attention to the relationships your numbers express when you make a calculation.) These two ideas go hand in hand. If you notice, for example, that every time you multiply a two-digit number by a one-digit number, you break up the two-digit number and add the pieces together, then you're one step closer to being able to use symbols to express the distributive property (which is a property of multiplication and addition that you probably use without even knowing it).

For example, if you can find the product 17×6 by thinking about 10×6 (which is 60) and 7×6 (which is 42), then you can write your computation this way: $17 \times 6 = 10 \times 6 + 7 \times 6$, which is an example of the distributive property: $(a + b) \times c = a \times c + b \times c$ (see Chapter 8 for more on the distributive property).

Some people define *algebra* as generalized arithmetic. Even and odd numbers are things you learn about when you're studying arithmetic. An *even number* is a whole number that you can share equally between two people without leftovers or fractions. (On the other hand, an *odd number* can't be shared equally between two people unless you have leftovers or fractions.) Children notice early in elementary school that the even numbers are those that you say when you count by twos: 2, 4, 6, 8, 10, and so on. Many children notice that adding two even numbers always gives an even sum — $2 + 2 = 4$ and $12 + 8 = 20$, for example. Adding two odd numbers also gives an even sum — $3 + 5 = 8$ and $11 + 7 = 18$. Similarly, the products of even numbers and of odd numbers also have patterns. These are all arithmetic ideas.

These sections describe the relationships between arithmetic and algebra in more detail.

Generalizing: The beginnings of algebra

Algebra begins when you try to study the structure of *all* even numbers and of *all* odd numbers. Mathematicians call this process of looking beyond examples to try to understand all cases, *generalizing*. When you generalize, you ask the question, "Is this always true?" One way to think about algebra is as *generalized arithmetic*, which means that you're trying to answer questions about all numbers, all addition problems, all multiplication problems, and so on.

For example, $4 + 6 = 10$ is an arithmetic statement. When you notice that 4, 6, and 10 are all even, you can generalize by asking whether the sum of any two even numbers is also even. Algebra helps you answer this question. Because even numbers of things can be shared equally between two people, you can write an expression such as $2a$ to represent all even numbers. Think of a as the (whole) number of things each person gets, then $2a$ represents both people's shares combined, which is an even number. If you try to share an odd number of things between two people, you have a leftover. An expression such as $2a + 1$ represents all odd numbers. Now imagine you have two different odd numbers: $2a + 1$ and $2b + 1$. Add those together and you have $2(a + b + 1)$.

The expression in the parentheses — $a + b + 1$ — is a whole number, so when you multiply it by two, the product is even. The algebraic symbols confirm what children suspected must be true — the sum of two odd numbers is even — which is true for *all* odd numbers.

Fifth graders may not create algebraic proofs such as this example, but they frequently express their computations in terms of the structure, not just the answer.

Determine how many bunnies are in the picture in Figure 10-1. More importantly, how do you know how many there are? One common way is to notice that each box has ten bunnies, so there are three groups of ten — 30 all together. Students may notate it as 3×10. But there are several other interesting ways to see these bunnies. For example:

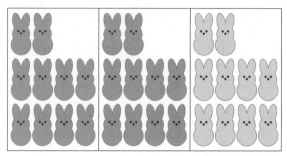

© *John Wiley & Sons, Inc.*

Figure 10-1: You can count these bunnies in many ways.

✔ Two rows of 12 and one row of 6: $(2 \times 12) + (1 \times 6)$

✔ Three boxes of 12, with two missing from each box: $(3 \times 12) - (3 \times 2)$

✔ Six columns of three and six columns of 2: $(6 \times 3) + (6 \times 2)$

All of these expressions have the same value; they all total 30, so they're all correct. The important thing for developing a sense of algebra is to pay attention to the structure of the computations and to compare these structures to each other. If students are accustomed to being sure that 3×10 and $(3 \times 12) - (3 \times 2)$ are the same as each other, then $a \times (b - c) = (a \times b) - (a \times c)$ won't be so scary in later grades.

Please Excuse My Dear Aunt Sally: Order of operations

If mathematics were something you (or your child) just did in your head or for your own purposes, you (or anyone else) wouldn't need an order of operations. But math isn't just a personal pursuit; it's a way that people communicate ideas with each other. Because your child uses math symbols to communicate, she needs to have some agreements with other people about what the symbols mean. The *order of operations,* or the sequence to follow when calculating the result of using different operations, is particularly significant for fifth graders.

Your child may use the mnemonic device *Please Excuse My Dear Aunt Sally*, abbreviated frequently as PEMDAS. Someone thought up this little tool long ago to help students remember the order of operations. PEMDAS stands for the following, which you do in this order:

- ✔ **Parentheses:** Do any of the computations inside of parentheses first (and follow the order of operations within those parentheses, if there is more than one operation inside them).

- ✔ **Exponents:** Calculate the results of any exponents or radicals (such as square roots).

- ✔ **Multiplication and division:** Do any multiplication and division next. These two operations come in the same step together, not one before the other.

- ✔ **Addition and subtraction:** Like with multiplication and division, do these operations in the same step.

Within each of these four steps, you go from left-to-right. So $8 - 4 - 1$ is 3, not 5. (You get 3 by subtracting 4 from 8 first and then subtracting 1.) In fact, a simple expression such as $8 - 4 - 1$ points out the need for an order of operations. It would be a disaster if people couldn't agree on the value of such a simple expression.

As an example of the disasters that can occur when the order of operations is unclear, you can easily find arguments about this expression on social media: $48 \div 2(9 + 3)$.

Following the strict PEMDAS interpretation, calculate the value of this expression like this:

1. **Add 9 and 3, leaving $48 \div 2(12)$.**

2. **Divide 48 by 2, leaving $24(12)$.**

3. **Multiply $24(12)$, getting 288.**

But PEMDAS misses some subtleties. Among these subtleties are the following rules, which are often unspoken:

- ✔ **Do implied multiplications before division or explicit multiplication.** *Implied multiplication* refers to something such as $2(3)$, where there is no multiplication symbol.

✔ **Factorials come before exponents.** A *factorial* is a number with an exclamation point after it (like this: 3!); they are useful in a number of areas in math, especially probability. The expression "3!" doesn't mean someone is exclaiming the number three; it means to multiply three by all the smaller whole numbers, ending with 1. So 4! means $4 \times 3 \times 2 \times 1 = 24$. PEMDAS doesn't have an F, so mathematicians have agreed that factorials come between parentheses and exponents in the order of operations, but this part of the rule isn't taught in school.

✔ **Calculate exponents within exponents from right to left.** An expression such as 3^{4^2} should be evaluated from the rightmost (or highest) exponent to the left (or down). So $3^{4^2} = 3^{16}$, not 81^2. (The difference between these two expressions is more than 43 million.)

The expression $48 \div 2(9+3)$ trips up people who pay attention to the unspoken rule about implied multiplication. These people will calculate the expression this way:

1. **Add 9 and 3, leaving 48 ÷ 2(12).**

2. **Multiply 2 by 12, leaving 48 ÷ 24.**

3. **Divide, getting 2.**

Or maybe it *doesn't* trip these people. Maybe these people got it right. There is really no way to know what the person who wrote the expression meant for it to say, and that brings me back to where I started.

The order of operations is an agreement among people that eases communication. If someone writes an ambiguous expression, such as $48 \div 2(9+3)$, then shame on that person for writing an ambiguous expression. If that person meant $(48 \div 2)(9+3)$, then that's what he should have written. If he meant $\frac{48}{2(9+3)}$, then he should have written *that*. The order of operations doesn't exist to trick people with ambiguous mathematical expressions; it exists so that people can communicate their ideas more effectively.

In any case, fifth graders learn the basic order of operations. They may or may not learn the mnemonic device about Aunt Sally. What matters is an agreement about order that makes it possible to communicate with each other. The order may not seem so important in elementary school when nearly all expressions have numerical values, but it becomes very important in algebra when expressions such as $3x + x$ are common and would have very different values under different assumptions about the order of operations.

Extending Place Value and Algorithms

In fifth grade, students study decimal place value and they master the third of four standard algorithms, which I discuss in the following sections. Both of these things build on students' prior learning about whole number place value.

Placing value: Left and right of the decimal point

In the earlier elementary grades, students mastered whole-number place value (refer to Chapter 7). Now in fifth grade, they study *decimal place value* — that is, they study digits to the right of the decimal point.

One of the most challenging things about studying decimal place value is that several of the ways of thinking about whole numbers that children commonly adopt don't apply to numbers to the right of the decimal point. Here are two examples where decimals behave differently from the expectations many students have based on their experiences with whole numbers. (I offer these examples to illustrate why it's important not to focus on rules, and why it's important instead to focus on meaning.)

> ✔ **For whole numbers, the number with more digits is larger. With decimals, the number with more digits may be smaller.** When you compare two whole numbers — say, 235 and 62 — the one with more digits is larger. That's because the 2 in 235 is in the hundreds place, while 62 has no hundreds at all. But with decimals, the story is different. 0.235 isn't larger than 0.62. Every time you add a digit to the right of a number, you add a *smaller* place value — this is the opposite of the process when you add digits to the left of the decimal point.

Students need to work in ways that keep their focus on the meaning of the places they work with, not just on generating new rules for this new environment. If students form bad habits with decimals, these habits prove remarkably robust and difficult to overcome. You will likely see your fifth grader decomposing decimals (for example, writing 0.62 as $\frac{6}{10} + \frac{2}{100}$) and drawing pictures in order to focus on the meanings of the numbers they work with rather than on memorizing a new set of rules.

✔ **In order to multiply by 10, you can add a zero on the end of the whole number. When you multiply decimals by 10, the digits move one place to the left.** When you multiply 42 by 10, you get 420 — the same digits, but with a 0 on the end. This fact makes multiplication by 10 much easier than multiplication by (for example) 9. It's an important rule that makes multiplication algorithms possible. But it isn't true for numbers to the right of the decimal point. 3.2×10 isn't equal to 3.20.

Multiplying decimals by 10 moves the digits one place to the left. In 3.2×10, the 3 moves from the ones place to the tens place and the 2 moves from the tenths place to the ones place to be 32. This is actually the same thing that goes on when you multiply whole numbers too. When you multiply 42×10, the 4 moves from the tens place to the hundreds place and the 2 moves from the ones place to the tens place to be 420. But then the ones place is empty, so you fill that place with a 0. As always in the Common Core Standards, the focus is on the meaning of the numbers, not on learning and applying rules without meaning.

As your fifth grader encounters decimals, help him pay attention to the meaning of the work that he's doing. Avoid formulating a bunch of new rules to remember, some of which may contradict the rules he learned for working with whole numbers.

Standard algorithms: Doing things the old-fashioned way

The first thing to know about the Common Core State Standards and the standard algorithms is that these algorithms are required. (*Standard algorithms* are nothing other than the ways of adding, subtracting, multiplying, and dividing multi-digit numbers that most people have learned in school for many years.) Students do need to learn to add, subtract, multiply, and divide in the way that you probably did in elementary and middle school, and the way your parents did if they went to school in this country, and so on.

Let me say again: The Common Core Standards require teaching standard algorithms. In fourth grade, students master the standard algorithms for addition and subtraction. In fifth grade, they're expected to master the standard algorithm for multiplication, and in sixth grade, they need to master the standard algorithm for division — what you may know as *long division.*

In the case of each operation, the standard algorithm comes at the end of the course of study. Students work with addition and subtraction in several grades before they're expected to master the standard algorithms, for example. Students are expected to be able to add and subtract multi-digit numbers fluently before fourth grade, but that's when they need to master the standard algorithms.

Standard algorithms such as multiplication are prized because they're efficient. Students must remember a small number of different steps that require little thinking as they execute them. Having a small number of different steps and requiring little thought are good when you want to compute quickly. That's why it's good to know standard algorithms. But you don't learn anything about place value or number relationships if you aren't thinking. That's why students learn standard algorithms last — after they have developed their understanding of place value and number relationships.

For example, a common error for students to make when using the standard algorithm for subtraction shows up in a problem such as 42 – 16. When the student notices that the 6 in the ones place of 16 is greater than the 2 in the ones place of 42, she just subtracts in the opposite direction: 6 – 2. Such a student gets an answer of 34 when subtracting 42 – 16.

Teachers need to help students find and fix such common errors. If the only tool a teacher and a student have in their mathematical toolkit is the standard algorithm, then the explanation of how to fix the error is complicated. "You need to cross out the 4 in 42 and write a 1 next to the 2 in 42, then think of that 1 and that 2 as 12. Subtract 6 from 12," and so on Mathematics presented this way has little meaning; it seems silly and arbitrary to many students.

If teachers and students have other ways of thinking about subtraction, thing can go much differently. "Imagine that you have 16 things, and you want 42 things. How many more would you need to get?" A child might be able to think, "I need 4 more to have 20, and then 22 more than that, so 4 and 22 is 26. I need 26 more things, so $42 - 16 = 26$." There are many ways to think this through. Children should be allowed and encouraged to do so.

The point is that an insistence on staying with the old-fashioned way — the insistence on a single way of thinking — makes it very difficult to help children when they're making consistent errors. When you're stuck in a rut, a fresh perspective is often helpful. Developing a variety of strategies for the operations is one way of providing that fresh perspective in the classroom and when students are out in the real world and need to work out a computation without the teacher looking on to catch their errors.

Operating on Fractions

Fractions are numbers. Students have studied the operations of addition, subtraction, multiplication, and division with whole numbers. In fifth grade, they study these operations on fractions, which these sections explain in greater detail.

Thinking about fractions

As students move from the fraction work of earlier grades (see, for example, Chapter 9), which involves thinking about the values of individual fractions and comparing one fraction to another, they begin to think about fractions in several different ways. As an adult, you may shift your own thinking quite easily among these different ideas, but they require a lot of work to learn the first time around. Those ideas include the following:

- ✔ **Part-whole:** This is probably the most common thing that comes to mind. In a part-whole model, when you see $\frac{4}{5}$, you may think about a circle that is cut into five equal pieces with four of them shaded.

- ✔ **Measurement:** Thinking about fractions on a number line is like thinking about measurements on a ruler. If you think of $\frac{4}{5}$ as a location on the number line and as a distance as well, then you're using a measurement idea for fractions.

- ✔ **Sharing:** You can think of $\frac{4}{5}$ as the result of sharing four cookies equally and completely among five people. Each person gets $\frac{4}{5}$ of a cookie.

- ✔ **Set:** Maybe you see five marbles — four of them are blue and one is white. You can say that $\frac{4}{5}$ of the marbles are blue.

These ideas about fractions (and others) are important for fifth graders to think about. They meet a variety of situations where fractions apply — not just shading circles — because fractions are useful for describing many different mathematical and real-life situations.

Adding, subtracting, and multiplying fractions

Students add, subtract, and multiply fractions in fifth grade. These skills require using and extending their knowledge of these operations from their work with whole numbers. The following sections explain in greater detail.

Adding and subtracting fractions

When adding two numbers, fifth graders need to express the same units. Three miles plus four gallons doesn't give you seven of anything because the units are different, which is true with adding fractions, too. When students get to fifth grade, they have a lot of experience with *unit fractions* (see Chapter 9). A *unit fraction* is a fraction with a 1 in the numerator, such as $\frac{1}{2}$ or $\frac{1}{12}$.

The *unit* in *unit fraction* refers to the 1 in the numerator. But a unit fraction is also a unit just like other units of measurement. A mile is a unit, a gallon is a unit, and a fourth is a unit. It's helpful for students to think about $\frac{3}{4}$ as three-fourths, just like they think of 3 miles and 3 gallons. This can remind them that $\frac{3}{4} + \frac{2}{4} = \frac{5}{4}$, and — more generally — that adding fractions requires common denominators because you need to add same-sized units.

Multiplying fractions

The hard part about multiplying fractions isn't the algorithm. It's an easy algorithm: Just multiply the numerators and multiply the denominators: $\frac{2}{3} \times \frac{4}{5} = \frac{2 \times 4}{3 \times 5}$.

Nor is the idea behind fraction multiplication particularly difficult to understand. When you multiply two fractions, you're finding parts of parts. When you multiply a fraction by a whole number, you're finding part of that whole number — half of 12, $\frac{1}{3}$ of 9 , $\frac{2}{3}$ of $\frac{4}{5}$. Students draw pictures to solve problems like in Figure 10-2 that involve parts of parts. You can see the doubly shaded $\frac{8}{15}$ in the upper right of the diagram. This product refers to a fraction of the original whole — the square in Figure 10-2.

Why fraction multiplication is called multiplication

The hard part about multiplying fractions is understanding why it's correct to call it *multiplication*. When you multiply whole numbers, you make multiple groups of the same size (as I discuss in Chapter 8). But when you multiply a fraction by something, you're finding *part* of a group and not multiple copies of the same group. So why should it be called multiplication? That's the difficult question here.

Mathematicians would say that the reason to call this multiplication is that it is *consistent* with multiplication of whole numbers. If you rewrite

3×4 as a fraction multiplication problem and multiply the numerators and denominators, you get $\frac{3}{1} \times \frac{4}{1} = \frac{3 \times 4}{1 \times 1}$, which is still 12.

Students explore multiple meanings for the process of multiplying numerators and multiplying denominators, including areas of rectangles and scaling; they think about $\frac{1}{3} \times 12$ as making 12 smaller by a factor of 3. In all of these cases, the idea of multiplication is related to the operation defined by multiplying the numerators and multiplying the denominators.

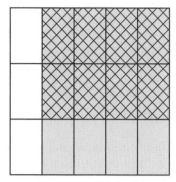

© John Wiley & Sons, Inc.

Figure 10-2: Finding two-thirds of four-fifths.

Measuring Volume and Graphing Data

Fifth graders continue to develop their understanding of measurement by studying volume and of data analysis by making line plots to represent the data they collect, which these sections explain.

Calculating volume: The number of cubes that fit inside

Volume is a lot like area. When students began the study of area in third grade (refer to Chapter 8), they find the number of unit squares that fit inside a rectangle and then develop a formula and extend the ideas to triangles, parallelograms, and other shapes in later grades. The same development applies to volume in fifth grade.

Students begin their work with volume by finding the number of unit cubes that fit inside a *rectangular prism* (this is the fancy word for a basic rectangular box). The *volume* is the product of the length, width, and height of the prism — $V = l \times w \times h$. If you think of the *base* (bottom) of the prism as a rectangle with dimensions l and w, then you can write the area of the base (which is the same as the number of cubes that would cover one layer on the bottom) as $B = l \times w$ and so $V = B \times h$.

In fifth grade, volume is limited to whole-number side lengths and rectangular prisms. More-complicated volume relationships then build on this foundation. (Refer to Chapter 13 for an example.)

Using a quick and dirty graph: The line plot

In fifth grade, students develop a simple data representation — the line plot. A *line plot* is basically a bar graph you can make as you collect your data. A line plot is built on top of a line that is marked off with the full range of possible data values. Then each time a value appears, you mark an X above that value, taking care to make the Xs the same size and to stack them neatly.

Figure 10-3 shows a line plot and a corresponding bar graph from a fictional fifth-grade class. Line plots allow students to quickly develop a way to visualize their data.

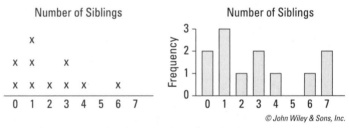

© John Wiley & Sons, Inc.

Figure 10-3: A line plot (left) and bar graph (right).

Concentrating on Properties of Shapes

Fifth graders develop their work with classifying shapes by taking on large categories of shapes and expressing the relationships among them. For example, students may look at all *quadrilaterals* (four-sided figures), sort them as squares, rectangles, parallelograms, rhombuses, and trapezoids, and then decide whether all squares are rectangles, all rectangles are parallelograms, and so on. A complete collection of these relationships is referred to as a *hierarchy* of quadrilaterals.

Part III

Moving Up to Middle and High School Math: Sixth through Twelfth Grade

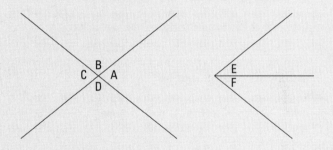

In this part . . .

✔ Read about the math that your middle school or high school child is studying in school this year in order to be better equipped to support your child's learning at home.

✔ Look at how the arithmetic of elementary school develops into the algebra of middle school and high school so that you can help your child navigate the transition to algebra.

✔ Understand the different possible structures for high school math programs so that you can better understand your child's path to high school graduation.

✔ Read about the biggest topics in middle school and high school — algebra, geometry, measurement, statistics, and probability — so that you can talk with your child about what she is studying.

Chapter 11

Relating to Ratios in Sixth-Grade Math

In This Chapter

▶ Grasping what ratios are and how they're different from fractions

▶ Solving ratio problems by thinking like a sixth grader

▶ Comprehending division of fractions

▶ Using rectangles to find areas of other shapes

Sixth grade is a transition year. Physically, many sixth graders are transitioning from elementary to middle school and from childhood to adolescence (at varying rates and starting times, of course). It's also a year of mathematical transition, too. During the elementary grades, the focus was on addition and subtraction relationships. In sixth grade, the focus shifts to multiplication and division relationships.

Students enter sixth grade already knowing how to multiply and divide, but they're more likely to think about comparing numbers, for example, using addition and subtraction than using multiplication and division. In sixth grade, students make this transition. The main tool for doing so is the idea of ratio.

In addition to studying ratios, sixth graders study a little bit of algebra, they work on areas of polygons by comparing to the area of a rectangle, and they begin to study more sophisticated data analysis tools. This chapter has you covered with these sixth-grade math concepts.

Understanding Ratios

A *ratio* is a comparison of two numbers that depends on the multiplication relationship between them. Here are some examples of ratios:

- A recipe for orange juice calls for 3 cans of cold water for each 1 can of frozen concentrate.

- The city of Saint Paul requires 5 parking spaces for every 1,000 square feet of retail space in a new development.

- A certain shade of green paint requires 3 parts of blue paint for every 2 parts of yellow paint.

In sixth grade, students study ratios that are set in understandable contexts, and they work these ratios in order to ask and to answer interesting and challenging questions.

The following sections help you know how ratios, rates, fractions, and proportions are related to each other. These sections also give you tips for working with these ideas.

Telling fractions from ratios

A *ratio* compares two numbers by using the multiplication relationship between them. You can write a ratio in three different ways:

- 3 to 2

- 3:2

- $\frac{3}{2}$

In each case, you can read the ratio as *three to two*. Sixth graders typically work with the first two of these forms rather than the fraction form because of important, but subtle, differences between fractions and ratios.

You can think of a fraction as a number. The fraction $\frac{3}{2}$ has a place on the number line — halfway between 1 and 2 (see Figure 11-1).

A ratio, though, isn't a number. It's a *comparison* of two numbers. A ratio doesn't belong on the number line. The notations *3 to 2* and *3:2* can help students remember that a ratio isn't a number.

© John Wiley & Sons, Inc.

Figure 11-1: A fraction is a number on the number line.

The reason the fraction form exists is that in some ways, ratios behave exactly like fractions. Think about the ratio in a recipe for green paint: *3 parts of blue paint to 2 parts of yellow paint.* If you make that recipe twice, your ratio will be *6 parts blue paint to 4 parts yellow paint,* and you'll still have the same color of green. In that way, 3:2 is the same as 6:4.

Working with fractions is identical. The fraction $\frac{3}{2}$ is equivalent to the fraction $\frac{6}{4}$ because they occupy the same location on the number line.

You can write equivalent ratios in the same way that you write equivalent fractions, which makes the fraction notation useful.

Solving ratio and rate problems

Sixth graders solve a variety of problems using ratios. The words *ratio* and *rate* are both appropriate in sixth grade and can mostly be used interchangeably (see Chapter 12 for more about these terms). Examples of ratios and rates that students work with include

- ✔ Dollars per hour
- ✔ Miles per hour
- ✔ Dollars per pound
- ✔ Students per class

Sixth graders develop a variety of strategies, which the following two sections identify, for solving ratio and rate problems. For an example, consider this problem:

> The ratio of boys to girls in Ms. Wales's class is 3:2. If there are 30 students in class, how many are boys and how many are girls?

Ratio table

Students may solve this problem with a *ratio table*, as Figure 11-2 shows. In a ratio table, a student keeps track of different equivalent forms of the ratio. The student may double (or triple, and so on) each value in the ratio to get larger values or may halve (or cut in three, and so on) each value to get smaller values. No two ratio tables for the same problem need to look identical, but they do need to maintain the given ratio throughout.

Boys	3	6	9	18
Girls	2	4	6	12
Students	5	10	15	30

© John Wiley & Sons, Inc.

Figure 11-2: A ratio table.

Equivalent fractions

A student may notice that if there are 3 boys for every 2 girls, then $\frac{3}{5}$ of the students are boys. She may write the following series of fractions in order to figure out the right number of boys for a 30-student class. In this case, each fraction represents the part of the class that is boys:

$$\frac{3}{5} = \frac{6}{10} = \frac{18}{30}$$

The last fraction shows that 18 out of 30 students are boys in this scenario.

The important thing about ratios is that they are multiplication-based comparisons of two numbers. If you double both numbers in a ratio — say 3:2 becomes 6:4 — the multiplication relationship between the numbers stays the same. In this example, 6 is still $1\frac{1}{2}$ times as big as 4, just as 3 is $1\frac{1}{2}$ as big as 2.

Examining Multiplication

Sixth graders study multiplication relationships — in contrast to the addition and subtraction relationships that dominated their thinking in elementary school. They do this by studying ratios and by studying factors and multiples and the division of fractions. These sections examine these concepts.

Differentiating between GCFs and LCMs

In fourth grade, students learn about *factors* and *multiples,* two terms that many students continue to confuse for each other in sixth grade and beyond. See Chapter 9 for more information about factors, multiples, and their uses. Sixth graders apply their knowledge of factors and multiples to look for the greatest common factor of two numbers and the least common multiple of two numbers.

The *greatest common factor* (or *GCF*) of two numbers is the largest number that is a factor of both numbers. The first step in thinking about GCF is to think about *common factors.* For example, think about the numbers 4 and 10. The factors of 4 are 1, 2, and 4. The factors of 10 are 1, 2, 5, and 10. The common factors of these numbers are 1 and 2. The *greatest* common factor is 2 because it's the biggest number on the list of common factors. So you can say that the GCF of 4 and 10 is 2.

The *least common multiple (LCM)* is the smallest (or *least*) number that is a multiple of both numbers. The first step in thinking about LCM is to think about *common multiples.* For example, think again about the numbers 4 and 10. The multiples of 4 are 4, 8, 12, 16, 20, 24, and so on (notice that these are the numbers you say when you skip count by 4, starting at 4 — *skip counting* is another way to think about the multiples of a number). The multiples of 10 are 10, 20, 30, 40, and so on. If you continue both of these lists far enough, there turn out to be many numbers on both lists: 20, 40, 60, and 80 are all examples of common multiples of 4 and 10. The *least* common multiple is the smallest number on that list: 20. So you can say that the LCM of 4 and 10 is 20.

Ask your sixth grader why there is no least common factor (LCF) or greatest common multiple (GCM). The answer is that the least common factor isn't interesting — 1 is a factor of every number so the LCF of two numbers would always be 1. The GCM, however, doesn't exist. All multiples of the least common multiple of two numbers are common multiples. For example, common multiples of 4 and 10 include 20, 40, 60, 80, 100, and so on. The list never ends, so no greatest common multiple exists (knowing this isn't a requirement of the Common Core standards, but it does make for a fun conversation).

Dividing fractions can be fun

In sixth grade, students learn to divide fractions. Maybe you learned the old line in school: "Yours is not to question why; yours is to invert and multiply." In a Common Core classroom, though, students will *need* to question why. The following sections give you a whirlwind tour of how sixth graders are likely to make sense of dividing fractions.

Comprehending the challenge of dividing fractions

You can think about division in two ways:

- **Sharing:** *Sharing division* tells you how many of something is in each share. For example, you may have 12 cookies that you want to share equally among 4 people, and you're wondering how many cookies each person can have.

- **Measuring:** *Measuring division* tells you how many groups you can make if you know the size of the group you want to make. An example is if you have 12 cookies and you want to put 4 cookies on each plate; you're wondering how many plates you can fill.

Sixth-grade students don't need to learn the language of *sharing* and *measuring*, but they should have had plenty of practice with both types in earlier grades. In sixth grade, they again work on both types of problems. For many students, measuring division problems are easier to think about than sharing division problems because sharing fraction division problems involve parts of a group, and students aren't accustomed to thinking about ideas such as "3 cookies is half a share." You may find it more challenging to think about sharing fraction division problems than measurement ones.

Either way, your time will be well-spent making sense of both. Chapter 8 provides more details about these two methods.

Solve the following problems by thinking through the relationships involved. Use the following sections to help you. (Don't go straight to the algorithm you learned in school; take a moment to think about these problems.)

> **Problem 1:** Griffin is baking cookies. He has 2 pounds of butter. Each batch of cookies requires $\frac{3}{4}$ of a pound of butter. How many batches of cookies can he make?

> **Problem 2:** Tabitha is riding her bike around the lake. She has ridden 2 miles, which is $\frac{3}{4}$ of the way around the lake. What is the total distance around the lake?

Problem 1 is a measuring division problem. You know how big each group (of butter) is, and you want to know how many groups that Griffin can make. Problem 2 is a sharing division problem. You know how many groups you have ($\frac{3}{4}$); you need to find the size of one group.

These two problems make most people think differently about division, even though both problems have the same numbers.

Finding a different algorithm

In solving Problem 1, most people think something like this: "There are eight fourths in two pounds, so I can make two groups of three-fourths, and I'll have two-fourths left over." Following this thinking, you can write $2 \div \frac{3}{4} = 2$, with a remainder of $\frac{2}{4}$. That remainder though is *two* of the *three* fourths of a pound that Griffin needs to make his third batch of cookies, so more precisely, $2 \div \frac{3}{4} = 2\frac{2}{3}$.

Often a picture helps to show this thinking. In Figure 11-3, each big square represents 1 pound of butter. The circled parts represent groups of $\frac{3}{4}$ of a pound. Notice that the third group has only two of the three parts needed to make a whole group.

One thing is missing from this equation: $2 \div \frac{3}{4} = 2\frac{2}{3}$. Eight fourths isn't in there. If you want to capture all of the thinking shown in the pictured strategy, you need to write $\frac{8}{4} \div \frac{3}{4} = 8 \div 3 = 2\frac{2}{3}$. This equation asks the question, "How many groups of 3 (fourths) are in 8 (fourths)?" Or more simply, "How many groups of 3 are in 8?" The picture doesn't appear to have either 8 or 3 in it, but they're there — the units

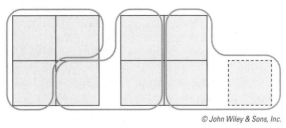

Figure 11-3: Dividing a whole number by a fraction.

have just been renamed. You're counting *fourths of a pound* instead of *whole pounds*. This way of thinking is a concrete application of the work students did naming and renaming units. (See Chapter 7 for more information about children's early experiences with units.) This kind of thinking leads to the *common denominator algorithm* for dividing fractions:

$$\frac{a}{b} \div \frac{c}{b} = a \div c$$

In the common denominator algorithm, you make sure that your fractions have a common denominator — the same as if you were going to add them (as I explain in Chapter 10). Then you divide the numerators. The common denominators ensure that the number you're dividing has the same units as the number you're dividing by — fourths by fourths, for example. In the example for Problem 1, you divide by three-fourths, so you wanted to write 2 whole pounds of butter in terms of fourths. There are 4 fourths in 1 pound, thus 8 fourths in 2 pounds. So the question becomes "how many groups of three (fourths) are in 8 (fourths)?" After things are measured using the same size unit, you only have to worry about how many there are, not their size.

Then, using the meaning of measuring division, the numerator of the divisor tells you how big each group is, so you divide the two numerators to find the number of groups you can make.

Questioning why you invert and multiply

In order to understand the more conventional algorithm known simply as *invert and multiply*, think about *sharing* division and reciprocals. The *reciprocal* of a fraction has the numerator and denominator swapped for each other. For example, $\frac{3}{4}$ and $\frac{4}{3}$ are reciprocals; so are $\frac{2}{99}$ and $\frac{99}{2}$. Deceptively, $\frac{1}{2}$ and 2 are reciprocals because $\frac{2}{1} = 2$.

Writing reciprocals is the *invert* part of invert and multiply. To understand why you invert and then why you multiply (instead of divide), it's helpful to know how reciprocals relate to each other.

Look at the diagram in Figure 11-4 and write down all of the fractions that you can think of that relate to this picture.

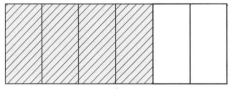

© John Wiley & Sons, Inc.

Figure 11-4: Identify the fractions that you see.

Maybe you see $\frac{4}{6}$ and $\frac{2}{3}$. Be specific about the meanings of fractions, so you could say that $\frac{4}{6}$ of the rectangle is shaded or that $\frac{2}{3}$ of the rectangle is shaded. These fractions compare the shaded part of the rectangle to the whole rectangle.

But what about the other way around? How does the whole rectangle relate to the shaded part? You can say that the whole rectangle is $\frac{6}{4}$ of the shaded part because the shaded part is made of four equal pieces, but the rectangle is made of six of these pieces.

This relationship will always be true. If x is a fraction (say $\frac{a}{b}$) of y, then y is the reciprocal fraction ($\frac{b}{a}$) of x. If you put it together with the idea that multiplication of fractions is scaling (that is, it makes numbers bigger or smaller; refer to Chapter 10 for more on scaling), then you can see that

$$\left(y \cdot \frac{a}{b}\right) \cdot \frac{b}{a} = y \cdot$$

You know at least one more thing about multiplication and division: If you multiply by a number and then divide by the same number, you get back where you started. For example, $5 \times 3 = 15$. Then $15 \div 3$ must be 5 because of the relationship between multiplication and division. Therefore, $\left(y \cdot \frac{a}{b}\right) \div \frac{a}{b} = y$.

To put all of this together, compare these two equations.

$$\left(y \cdot \tfrac{a}{b}\right) \cdot \tfrac{b}{a} = y$$

$$\left(y \cdot \tfrac{a}{b}\right) \div \tfrac{a}{b} = y$$

If you replace $y \cdot \tfrac{a}{b}$ with x and set the two expressions equal, you have $x \div \tfrac{a}{b} = x \cdot \tfrac{b}{a}$. That is, *invert and multiply*.

Sixth graders are much more likely to make the argument that leads to the common denominator argument than they are to make the much more abstract one that leads to *invert and multiply*.

Extending from Arithmetic to Algebra

Sixth grade is the first time that students are required to use letters that stand for numbers. Sixth graders typically use variables that closely match the values they represent, so s for *side length,* v for *volume,* and so on. Also, they notice that some expressions — even though they may look quite different — always have the same values as each other, and they explore this property, called *equivalence.* These sections take a closer look at variables and equivalence.

Using variables

Sixth graders are introduced to variables for the first time. Mostly, they're concrete applications of variables in which students represent simple numerical or measurement relationships.

An example of using variables to represent numerical relationships might be expressing the number of wheels on x cars as $4x$ or an unknown number of boys in a class as b, an unknown number of girls in a class as g, and the number of students in the class as $b + g$.

An example of using variables to represent measurement relationships is showing the perimeter of a square as $4s$ and the area of a square as s^2. In these two cases, s represents the side length of the square, and students show how to use the side length to compute the other measure — either perimeter or area.

Beginning the use of variables by expressing these relationships in familiar scenarios allows students to get a feeling for the way algebra shows the structure of a situation, whereas arithmetic only shows individual values.

Figuring out equivalence

Students come to sixth grade knowing that there are many ways to write the same value. The number 459 can be written as $400 + 50 + 9$, for example, or $460 - 1$, or $9 \cdot 51$, and so on. Similarly, the fractions $\frac{1}{2}$, $\frac{2}{4}$, and $\frac{3}{6}$ all have the same value. Students are accustomed to calling fractions *equivalent* when they have the same value.

This idea of same value is extended in sixth grade to include equivalent expressions. Two *equivalent expressions* have the same value whenever their variable has the same value. The expressions $4(x + 2)$ and $4x + 8$ have the same value as each other for all values of x, so they're equivalent. The expressions $x + 2$ and $2x$ only have the same value when x is equal to 2, so they aren't equivalent.

In sixth grade, students use simple examples of the *distributive property* (this property states that $a \cdot (b + c) = a \cdot b + a \cdot c$ — if you think of $12 \cdot 7$ as $70 + 14$, then you're using the distributive property because $12 \cdot 7 = 10 \cdot 7 + 2 \cdot 7$) to write and to justify that two expressions are equivalent. The example of $4(x + 2)$ and $4x + 8$ is an example of the distributive property. (Head to Chapter 8 for more about this property.) Many sixth graders justify that two algebraic expressions are equivalent through a combination of checking a few sample values of the variable and gut instinct. In formal mathematics, examples and instinct aren't strong enough justification for knowing that something is always true. As a way to get started with thinking about algebraic expressions, they're highly valued.

Measuring Area and Volume

As students work with measurement in sixth grade, they find areas of a variety of polygons by relating to the areas of known figures — especially rectangles — and they get started with three-dimensional measurement by looking more deeply at volumes of rectangular prisms, as the following sections explain.

Focusing on area

You may remember from your days in school the study of area as being about long lists of unrelated formulas — one formula for triangles, another for parallelograms, and a really strange one for trapezoids. A major goal of the measurement strand in Common Core is for students to be able to relate these formulas to each other so that they can understand area and can remember (or figure out) these formulas when they are needed.

Area all goes back to rectangles. The area of a rectangle is equal to the product of the rectangle's width and length. That is, for a rectangle, $A = l \cdot w$. Imagine a grid of square units filling up the rectangle, and this imaginary grid has l rows with w square units in each row. In Chapter 8, I introduce the meaning of $l \cdot w$ as l groups of w. That meaning is important to the study of area.

A right triangle is half of a rectangle (refer to Figure 11-5), so for right triangles, $A = \frac{1}{2} l \cdot w$.

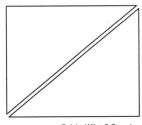

© John Wiley & Sons, Inc.

Figure 11-5: Two right triangles make a rectangle, so the area of each triangle is half the area of the rectangle.

For all triangles, not just right triangles, you can relate the area to the area of a rectangle. Typically, you use the terms *base* and *height* instead of *length* and *width*, so the formula for area of a triangle is $A = \frac{1}{2} b \cdot h$. The reason is that the *height* of a triangle usually isn't the same as a *side length*. In the case of a rectangle, the base and height are the same as the side lengths, so *base* and *height* are terms that always work, while *length* and *width* are only useful for rectangles.

You sometimes have to be clever though and cut one of the triangles apart to see that two triangles make a rectangle with the same base and height as the triangle, as Figure 11-6 shows.

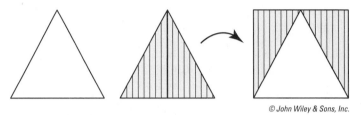

© John Wiley & Sons, Inc.

Figure 11-6: Two non-right triangles also make a rectangle.

Sixth graders also find the formula for area of a parallelogram based on formulas they already know. It may involve relating parallelograms to rectangles or knowing that two identical non-right triangles make a parallelogram as in Figure 11-7. In either case, they find that for parallelograms, $A = b \cdot h$.

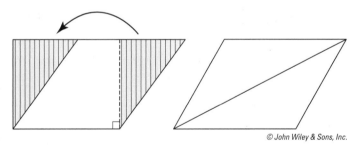

© John Wiley & Sons, Inc.

Figure 11-7: Making a parallelogram into a rectangle (left) or two triangles (right) to find its area.

Finally, sixth graders cut polygons into triangles and find the areas of the individual triangles in order to find the areas of polygons for which they don't have formulas.

The case of the tilty triangle

The tilty triangle is more complicated and requires either a tremendous amount of patience for cutting and pasting or some clever algebra manipulations. The following figure shows the triangle inside a rectangle. (There is no mathematically correct term for a *tilty* triangle — I'm just referring to a triangle whose top vertex is not directly above its base.) The area of the rectangle is $(b+a) \cdot h$. Two right triangles are inside the rectangle. The area of the one on the right is $\frac{1}{2}a \cdot h$, and the area of the one on the left is $\frac{1}{2}(a+b) \cdot h$. The tilty triangle has an area of $(a+b) \cdot h - \frac{1}{2}a \cdot h - \frac{1}{2}(a+b) \cdot h$. After a little bit of algebraic manipulation, this last expression shows that the area of even a tilty triangle is found with the expression $\frac{1}{2}b \cdot h$.

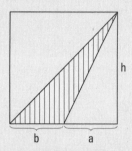

Sixth graders tend to work through the cases of right triangles and the usual non-right triangles fairly naturally. The case of the tilty triangle usually involves a lot more effort because it requires a bit of algebra work, which is fairly new to sixth graders.

Getting 3-D with volume

Sixth graders extend their understanding of volume by considering the case of fractional edge lengths. For example, you can't fill a 2-inch-by-2-inch-by-$1\frac{1}{2}$-inch rectangular prism with 1-inch cubes. You can fill it partway, but you'll have some leftover space. Students imagine filling this empty space either with partial cubes or with smaller cubes that have fractional side lengths. However they visualize it, the principle is the same as when the prism has whole-number side lengths for rectangular prisms: $V = l \cdot w \cdot h$ or $V = B \cdot h$, where B is the area of the bottom (or base) of the prism. This understanding is important for finding volumes of other prisms and of cylinders in later grades.

Measuring Datasets

Sixth grade is when students get serious about analyzing datasets in two ways, which these sections explain. These sections examine what students measure: data collections by using measures of center (such as mean and median) and measures of spread (such as range and the intimidatingly named mean absolute deviation). Together, these two measures can be useful in describing a set of data and in comparing two or more different sets of data.

Eyeing measures of center

Sixth graders use measures of center to answer questions about what is typical. You can use a measure of center to describe the height of the *typical* sixth grader and the height of the typical fifth grader. Then you can compare these values. A measure of center is a very crude representation of the whole collection of data. Sixth graders study three measures of center, which include the following:

- ✔ **Mean:** The *mean* is what you probably think of as *average*. You find the mean of a set of data by adding all of the values together and dividing by the total number of values in the set. For example, imagine that you had three children and that the youngest 1 is 42 inches tall, the middle child is 51 inches tall, and the oldest child is 72 inches tall. Then the mean of these heights is $\frac{42+51+72}{3} = 55$ inches.

- ✔ **Median:** The *median* is the middle value in a set of data that has been put in order. The median of the children's heights in the example is 51 inches. In the case of a dataset with an even number of data points, *two* values are in the middle. If those values are different from each other, the median is the number halfway between them.

- ✔ **Mode:** The *mode* is the most common value in a set of data. Although you can find the mode of a set of numerical data, the mode is especially important because it's the only measure of center that makes sense when your data doesn't consist of numbers. For example, if you ask everyone at school for their favorite color, you can't add these colors together, so you can't find the mean. Any order you would put the colors in is arbitrary, so you can't find the median. This leaves the most common color as the preferred measure of center — the mode. Another difference between mode and the other measures of center is that there can be two (or more) modes. If there is a tie between two favorite colors, the data is *bimodal*. There can also be no mode if everyone has a different favorite color.

For example, consider how you can use these measures. If you want to answer a question such as "Are sixth graders taller than fifth graders?" you need to collect some data. In this case, you would need to collect the heights of a large number of sixth graders. When you do, it will become clear that *many* sixth graders are taller than many fifth graders, but also that some fifth graders are taller than some sixth graders. The tallest kid possibly may be a fifth grader, but that doesn't mean that fifth graders as a group are taller than sixth graders. Instead, when you compare the *mean heights* of these two grades, you can see that sixth graders are a little bit taller than fifth graders. Similar results come from comparing the medians or modes of the height data.

Focusing on spread

Sixth graders study two measures of spread:

- ✔ **Range:** The *range* is simply the difference between the largest value and the smallest.

- ✔ **Mean absolute deviation (MAD):** *MAD* is more complicated. It describes the average difference between a value in the dataset and the mean.

- ✔ Consider the dataset of children's heights in the last section: 42, 50, and 72 inches. The mean of these values is 55 inches. The actual heights are different from the mean by 13, 5, and 17 inches, respectively. The mean of these values is $11\frac{2}{3}$, which means that the average height in this dataset is $11\frac{2}{3}$ inches away from the mean — maybe greater and maybe less.

- ✔ A dataset that has a large MAD is more spread out — the values differ more from the mean — which is why the MAD is called a measure of spread. In high school, students learn about *standard deviation*, which is the most common measure of spread used in statistics. (Chapter 14 discusses high school math.)

Chapter 12

Pursuing Proportions in Seventh-Grade Math

. . . .

In This Chapter

▶ Understanding the role of proportional relationships in seventh grade

▶ Exploring why a negative times a negative is a positive and other puzzles of integers

▶ Taking another step toward proof in geometry

▶ Grasping how probability and statistics play out in seventh grade

. . . .

*S*ome people describe the seventh-grade year as being about proportional reasoning. This is partly true: Fractions, ratios, rates, and proportions form a continued focus for the work of seventh grade. But a lot more is going on, too. As this chapter explains, students continue their work in geometry and measurement by figuring out whether triangles are congruent (and how to be sure they're right about that), and by studying circumference and area of circles. Seventh graders work with more complicated and interesting sets of data and data analysis techniques. They also study probability by asking how likely something is and then testing their answers by rolling dice, spinning spinners, and so on.

Examining Ratios and Proportional Relationships

Seventh graders compare numbers in two major ways:

- ✔ **With differences:** When you overdraw your bank account, your negative balance is the *difference* between the size of the check that you wrote and the amount of money you had in the account. You find a difference by subtracting one number from the other. *Five more than* is an addition relationship that describes a difference.

- ✔ **With ratios:** When you get lucky on a TV game show and double your money, the *ratio* of the money you have now to the money you had is 2:1 (read this as "two to one"). You find a ratio by thinking about the multiplication relationship between two numbers. *Twice as big* is a multiplication relationship that describes a ratio.

Throughout most of the elementary grades, children think about adding and subtracting when they compare. Students answer these kinds of comparison questions: "How many inches taller am I than you?," "How many more girls are in class than boys?," and "How much warmer is the weather today than it was yesterday?"

In middle school, students focus on comparing with ratios. "What happens to the perimeter of a shape when I double its side lengths?" "What happens to the area?" "If a 12-gallon tank of gas costs $39.59, how much does each gallon cost?" These questions involve ratios, rates, and proportional reasoning.

The proportional reasoning of seventh grade is an important step on a long journey to higher mathematics. The next sections help you understand the importance of proportions and give you insight on student solution methods.

Keeping track of key terms

Some vocabulary terms are especially important for your seventh grader. Here is a quick guide to the key terms in this area:

- ✔ **Ratio:** A *ratio* is a comparison of two numbers (usually both are non-zero). Usually, ratio is used for part-part comparisons, but not always. Chapter 11 discusses the several forms that you can use to write ratios.

✔ **Rate:** The word *rate* makes most people think about change. Most people use the term *ratio* in situations where the numbers don't change and use *rate* when the numbers are changing.

If there are five girls for every three boys in a class, it would be correct to call *five girls for every three boys* a *rate*, but *rate* sounds awkward to most people. It's better to use the term *ratio*. If students were enrolling in a school and five girls were enrolling for every three boys, the term *rate* is a more natural fit because the number of enrollees is increasing.

✔ **Unit rate:** A *unit rate* is a rate where one of the numbers being compared is 1. If five girls enroll for every three boys, this isn't a unit rate. The unit rate would be $\frac{5}{3}$ girls per boy enrolling at the school. The word *per* means *for each one*, which helps you notice the unit rate.

Seeing how proportions are key

The word proportion may make you think of cross-multiplying and dividing, but for many people, *cross-multiply* and *divide* are rules without reasons. When students develop their own strategies for solving proportions, they can use the same kinds of thinking to make progress in algebra.

Officially, a *proportion* is an equation with a fraction on either side of the equal sign. Figure 12-1 shows what proportions look like.

$$\frac{1}{3.69} = \frac{12}{x}$$

$$\frac{x}{8} = \frac{7}{1}$$

$$\frac{5}{40} = \frac{3}{x}$$

You use proportional reasoning every day, whether you think of it that way or not, such as these examples:

✔ How much will a full tank of gas cost if 1 gallon costs $3.69?

✔ I ran 5 miles in 40 minutes yesterday. Do I have time to run 3 miles right now?

✔ If I double my cookie recipe, how much flour will I need?

✔ My neighbor's dog is 8 years old. How old is she in dog years?

Seventh graders work these kinds of problems frequently in class. They make graphs and tables to show these relationships, and they write equations to better understand how numbers relate to each other proportionally.

The heart of all proportional relationships is the simple formula $y = kx$. Think of x as one variable (say the number of miles you run), and y as the other variable (say the number of minutes you run). Then k shows the relationship between them — the unit rate (in this example, k would be the number of minutes it takes you to run a mile on average).

A graph of $y = kx$ is straight line through the origin, and all the points on the line are possible values in the proportional relationship (say, possible numbers of miles and the corresponding times). A table for $y = kx$ shows that every x-value, when multiplied by k, gives the matching y-value as Figure 12-1 shows.

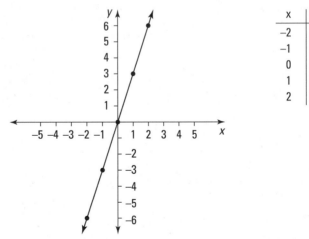

© John Wiley & Sons, Inc.

Figure 12-1: A table and a graph representing the same proportional relationship.

You may notice that k is an important value in proportional relationships. That is why k gets a special name — it's the *constant of proportionality*. You may better know the constant of proportionality as *slope* or *rate of change*.

Solving proportions

Developing skills for knowing how to solve proportions is important in the seventh grade. In a Common Core classroom, the emphasis is on making sense of the solution method, rather than on memorizing a given procedure. One general rule for whether your student is making sense of his solution is this: Does he know what each number means along the way?

Consider this question: I ran 5 miles in 40 minutes yesterday. If I run at the same rate, do I have time for a 3-mile run right now? In order to answer this question, a student could figure out the time it should take me to run 3 miles. He might write this proportion: $\frac{5}{40} = \frac{3}{x}$.

In the spirit of keeping track of the meaning of numbers as you work, here are two ways a typical seventh grader might solve this proportion. In each case, I discuss what meanings the student might have for the numbers:

- **Unit rate:** $\frac{5}{40}$ is equal to $\frac{1}{8}$. This means that I run 1 mile every 8 minutes (on average) if I ran 5 miles in 40 minutes. One mile every 8 minutes is a unit rate. (See the earlier "Keeping track of key terms" section for specific information about unit rates.) Using this unit rate, I must run 3 miles in 24 minutes.

  In this solution, every number has a meaning. The fraction $\frac{1}{8}$ is a rate that compares miles to minutes. You can think of this as "1 mile for every 8 minutes" or as "one-eighth of a mile per minute." Either way, you can multiply 8 by 3 because running 3 miles takes 3 times as long as running 1 mile. Because 8 • 3 is 24, it should take 24 minutes to run three miles. This strategy builds on strategies for working with equivalent fractions and ratios, which are already familiar to students.

- **Cross-multiply and divide:** Multiplying 40 by 3 gives 120. (This is the cross-multiplication part.) Dividing 120 by 5 gives 24, so I can run 3 miles in 24 minutes.

In this solution, it's not at all obvious what the 120 means in the context of running. Forty minutes times 3 miles equals 120-minute miles? This doesn't make sense. And then why divide 120 by 5? Without a meaning for the 120, it's hard to say why you're dividing.

A typical seventh grader can work out the first solution, given time to think. This solution makes sense, and it's based on other things that he knows well. A seventh grader would be unlikely to come up with the second solution alone.

In a Common Core classroom, teachers encourage students to develop strategies, such as the first solution, rather than telling students solution methods that don't make sense to them.

Working with Negative Numbers

In seventh grade, students add, subtract, multiply, and divide both positive and negative numbers. Addition and subtraction take place on the number line. The rules for multiplication and division depend on properties of numbers.

On the number line, numbers have two important meanings:

- ✔ A number is a *point* on the number line.
- ✔ A number is a *distance* on the number line.

What the first one means is that each little tick mark you put on a number line represents a number — 0, 1, 2, and so on: Each of these is matched with a point on the number line.

The number line is a useful tool for adding and subtracting numbers. Many people struggle to explain why $-7 - (-9) = 2$. But if you think about these numbers on the number line, $-7 - (-9)$ is asking "How far from -9 is -7?" Well, -7 is two spaces to the right of -9 on the number line, so that's the answer. (See Figure 12-2.)

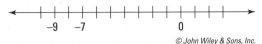

© John Wiley & Sons, Inc.

Figure 12-2: The distance between –9 and –7.

The answer to that example (called the *difference* because it's the answer to a subtraction problem) isn't a point on the number line, though. It's the distance between those two points. Actually, it's a *directed* distance, which means that keeping track of the direction you go is important when you move between –7 and –9. Moving to the right is positive; moving to the left is negative, as Figure 12-3 shows.

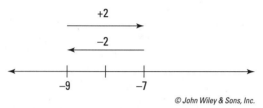

Figure 12-3: The directed distances between –9 and –7.

Similarly, you can use number lines to think about addition. $9 + 3$ means *start at 9, move three spaces to the right.* You don't need a number line to know that the sum is 12, but doesn't it help to know that this works out on the number line, too? This idea of *start + change = end* is more helpful with a problem such as $–6 + 8$. Start at –6, move 8 spaces to the right, and you end up at 2; refer to Figure 12-4.

So, $–4 + –7$ means start at –4 and move 7 spaces to the left. You end at –11, so $–4 + –7 = –11$.

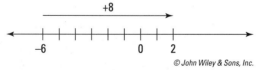

Figure 12-4: Moving 8 spaces to the right of –6.

Multiplying and dividing with properties

When you multiply a negative number by a negative number, you get a positive number. Why this is the case is perhaps the most perplexing question in all of mathematics. Mathematicians agreed that negative numbers needed to follow the same rules as positive numbers.

Let me explain. Numbers, multiplication, and addition have properties that are true of all positive numbers. When it came time to figure out how to work with negative numbers, a rule that mathematicians agreed on is that these same properties needed to be true when using negative numbers, too.

The terms *associative property* and *distributive property* sound fancy, but they refer to things that are intuitively true about numbers (for more details on these properties, see Chapter 8). The associative property is what you use when you multiply 80 • 4 in your head. You probably think to yourself, *8 times 4 is 32, and 32 times 10 is 320.* Formally, the associative property is $(a \cdot b) \cdot c = a \cdot (b \cdot c)$. So for this example, if a is 10, b is 8, and c is 4, then $(10 \cdot 8) \cdot 4 = 10 \cdot (8 \cdot 4)$.

The distributive property is what you probably use if you can multiply 13 • 9 in your head (and don't worry — no shame if you can't). Many people think this way: *10 times 9 is 90, and then 3 times 9 is 27, so 13 times 9 is 90 + 27, or 117.* Formally, the distributive property is $(a + b) \cdot c = a \cdot c + b \cdot c$.

It's really helpful if the same rules apply to multiplying negative numbers as apply to positive numbers. But — and this is important — it didn't have to be this way.

After mathematicians decided these same rules would apply, then a negative times a negative had to be a positive. Here's why. Consider this expression: $-3(1 + -1)$. Add inside the parentheses first, and you get $-3 \cdot 0$, which is equal to 0, so $-3(1 + -1) = 0$. But the distributive property tells you that you don't *have* to add inside the parentheses first. You can write $-3 \cdot 1 + -3 \cdot -1$. It has to be equal to 0 (because of the distributive property), so $-3 + (-3 \cdot -1) = 0$. Whatever $(-3 \cdot -1)$ is, you have to be able to add it to -3 to get 0. There is only one possibility: positive 3. That means $-3 \cdot -1$ has to be equal to positive 3.

Nothing was special about -3 in that example; things work out the same way whatever negative number you choose. The only thing special about -1 in the argument is that it's the opposite of 1. You can do the same thing with $(2 + -2)$ or any pair of opposites.

Describing Things with Algebra

When you study algebra you work with expressions and equations. An *expression* is a collection of mathematical symbols. An *equation* has an equal sign and states that two expressions are the same as each other. Often people think of an expression as being like a phrase: $5+3$ is an expression, as is $3x-2$. An equation, then, is like a sentence: $5+3=8$ is an equation, and $y=3x-2$ is also.

Expressions and equations are some of the building blocks of algebra. In a Common Core classroom, students use algebra to describe their understanding of the mathematical relationships in the world, to state patterns and structure that appear in their arithmetic, and to notice and state new relationships. In seventh grade, most of these relationships are *proportional*, which means that there is a constant rate of change in one variable relative to another.

Students' experiences with arithmetic and with real-world mathematical relationships (such as interest) are the basis for learning algebra in a Common Core classroom. This is a far cry from the decoding of complicated and meaningless word puzzles that has cluttered algebra classrooms for many years.

Story problems or *word problems* — it doesn't matter what you call them — have been the heart of algebra teaching for many years and a source of anxiety for many students. Problems, such as the following one, force students to ask when they are ever going to use algebra.

> Lucie bought 36 apples at the grocery store. She bought three times as many red apples as green apples, and twice as many green apples as yellow apples. How many yellow apples did she buy?

Seventh graders know that there is no possible scenario in which Lucie would know that she bought twice as many green apples as yellow apples without knowing how many of each she bought. Some students view these problems as fun little puzzles. But many students take these problems as a sign that algebra is not useful for much of anything.

A Common Core classroom has fewer of these silly word problems. Instead, algebra involves looking for the mathematical structure in everyday situations and in the arithmetic with which students are familiar. In short, students learn introductory algebra as an extension of what they already know.

For example, seventh graders already know how to find 1 percent of the money in a savings account (this is a realistic interest rate at the time of this writing) in order to calculate the value of the account after a year. Using their knowledge from elementary grades, students are likely to multiply .01 by the balance of the account, and then add the result to the balance.

Doing this with a $500 bank account goes like this: .01 • 500 = 5, then 500 + 5 = 505. Seventh graders use algebra in two ways to describe what is happening.

✔ **They notice that the procedure is the same no matter what the beginning balance is in the account.** Nothing was special about 500, so students can use any number and the procedures are the same. Students can write the general relationship using variables. In this case, they can write x for each instance of the number 500, like this: $x + .01x$.

✔ **When students use the distributive property to rewrite $x + .01x$ as $(1 + .01)x$, or $1.01x$, they're making an important algebraic leap.** (Refer to the earlier section in this chapter, "Multiplying and dividing with properties," as well as Chapter 8 for more information on the distributive property.) Writing $x + .01x$ makes it difficult to calculate *two* years worth of interest. When you try to write the expression things get complicated. But writing it as $1.01x$, each extra year just requires one more set of parentheses. Two years of interest looks like this: $1.01(1.01x)$ because x is the initial balance and $1.01x$ is the balance after one year, which means $1.01x$ is the initial balance for the second year and so on. This particular relationship forms the foundation for exponential functions in high school (check out Chapter 14 for more information), but the principle applies everywhere in the algebra curriculum.

Delving Deeper into Geometry

In seventh grade, students wrap up the measurement of two-dimensional shapes by finding formulas for perimeter and area of circles, and they take some important steps to studying proof in high school geometry by studying relationships among triangles. The following sections take a closer look at these two concepts.

Measuring circles: A piece of pi

Get yourself a circle. Measure its *diameter* (the distance from one side to the other of your circle, going through the center). Measure its *circumference* (the distance around the circle). If you're careful, you find that the circumference measurement is a bit more than three times the diameter measurement. That relationship is π (*pi*).

Seventh graders are likely to learn about π following the preceding calculation in a Common Core classroom. Students measure the diameter and circumference of many circles, and they investigate the relationship between these measurements. This relationship is proportional so it's a special case of the form $y = kx$ where C (for circumference) plays the role of y, d (for diameter) plays the role of x, and π is k. The standard formula for circumference of a circle, then, is $C = \pi d$

The *radius* of a circle is half its diameter; the radius is the distance from the center to the edge of the circle. That means you can also write $C = \pi 2r$ or the more conventional $C = 2\pi r$.

π is useful for finding the area of a circle too. Here's why. Cut your circle into wedges like slices of pie. You can rearrange those slices to make something that looks a bit like a parallelogram in Figure 12-5. The height of that (almost) parallelogram is the same as the radius of the circle. The base is the same as half the circumference (the other half of the circumference is along the top of the parallelogram). The area of a parallelogram is *base times height* (which I discuss in Chapter 11), so you have $A = r(\pi r)$, which is usually rewritten as $A = \pi r^2$.

Unlike circumference, area does *not* have a proportional relationship with the diameter or the radius. If the radius doubles, then the area more than doubles. $A = \pi r^2$ isn't in the form $y = kx$.

Tackling triangles

Viewed on their own, the seventh grade standards about angles and triangles can seem a little odd. In this section, I put these standards in context to help you understand why seventh graders study what they do.

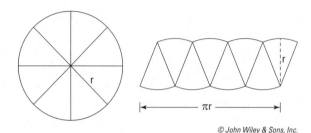

© John Wiley & Sons, Inc.

Figure 12-5: Rearranging parts of a circle to make a (near) parallelogram.

The road from elementary school geometry to the high school course called *geometry* has been rocky. Traditionally, children have spent a great deal of time recognizing and identifying shapes in the primary grades but have encountered a huge gap and advance little in their geometry knowledge until tenth grade, when they're expected to prove geometry theorems.

The Common Core standards seek to fill that gap by slowly making the geometry more challenging at every grade, which means for seventh graders that they're taking on a small part of what is traditionally in a high school geometry course. Seventh graders play around with drawing triangles in an attempt to determine precisely what conditions are necessary in order to know that two triangles are identical to each other (the math term for identical is *congruent*).

If the triangles have three pairs of congruent sides and three pairs of congruent angles, then the two triangles are congruent (refer to Figure 12-6 for an example). But the three angles of a triangle add up to 180°, so if you only know that two of the three pairs of angles are congruent, the third pair must be, which means that you only need three pairs of sides and two pairs of angles. Seventh graders build this kind of argument as they investigate which conditions must give congruent triangles and which may not.

Studying the properties of angles supports some of this work. A bunch of vocabulary goes along with it. Two angles are *supplementary* if their measures add to 180°. Two angles are *complementary* if their measures add to 90°. Two angles are *vertical* if they result from two intersecting lines but have only a vertex in common. Two angles are *adjacent* if they share a side. In Figure 12-7, angles A and B are supplementary. Angles E and F are complementary. Angles A and C are vertical. Angles A and B are adjacent (and so are E and F).

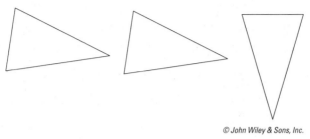

© John Wiley & Sons, Inc.

Figure 12-6: Congruent triangles.

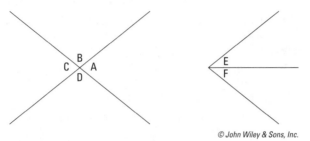

© John Wiley & Sons, Inc.

Figure 12-7: Supplementary, adjacent, and vertical angles.

Studying Data and Chance

In seventh grade, students work on basic probability ideas and on statistics that help them compare populations to each other. These sections examine the important ideas in seventh-grade statistics and probability.

Considering basic probability

Probability is the study of how likely something is to happen. In seventh grade, students develop theories about probability and then test these theories with experiments.

The common example is flipping a coin. If the coin isn't weighted, if you let it hit the ground, and if you don't otherwise interfere with the flip, then the probability of getting heads is $\frac{1}{2}$. The probability of getting tails is also $\frac{1}{2}$, so these two outcomes are equally likely.

But say that someone hands you a coin. You flip it three times and they're all heads. You flip it three more times and get heads again on all flips. At what point do you begin to suspect that there is something wrong with the coin? This is the kind of thinking seventh graders are doing in the Common Core.

Although they may not work with weighted (or otherwise unfair) coins, seventh graders do develop theories, make claims and then flip coins, spin spinners, draw blocks out of bags, and so on, in an effort to test these theories against reality. Just as with the coin example, an important question for seventh graders to consider is "How far off do my model and my results have to be before I start to rethink my model?"

Examples of the kinds of questions that can spark this model building include

- ✔ If I flip two coins, what is the probability that they both come up the same?
- ✔ If one in six bottles of pop is a winner in a contest, does that mean I am guaranteed to win if I buy a six-pack?
- ✔ What is the probability of rolling a sum of 3 when I roll two six-sided dice?

Making sense of data: Statistics

The process of drawing conclusions from incomplete information is called *statistical inference*. Seventh grade is where students get their first taste of statistical inference.

They may consider a question like this that requires statistical inference: "How do people know how many deer live in the state of Minnesota?" A certain number of deer are tagged each year; they're marked in some permanent way. Then people keep track of what fraction of the deer killed in hunts or in collisions with automobiles and so on are tagged. That fraction is approximately the same as the fraction of the original population that got tagged. So if 200 deer were tagged, and then $\frac{1}{100}$ of the deer killed in that year's hunting season were found to be tagged, then 200 is about $\frac{1}{100}$ of the population. There must be about 20,000 deer altogether. In this example, the incomplete information is the precise fraction of the population that is tagged.

Seventh graders don't use the full range of tools that statisticians use to do random sampling. But in drawing conclusions from incomplete information, they do develop the ways of thinking that are important in statistics.

Chapter 13

Arriving at Algebra in Eighth-Grade Math

. . . .

In This Chapter

▶ Wrapping up numbers

▶ Getting serious about algebra and functions

▶ Measuring new volumes

▶ Meeting the Pythagorean theorem

. . . .

*T*his chapter examines eighth-grade math, which focuses mostly on algebra. Eighth graders study the interesting case of irrational numbers and then shift their attention to algebra. They pay careful attention to linear relationships, learn about functions for the first time, and connect their algebraic work to their statistics knowledge. Students also learn, prove, and use the Pythagorean theorem and study volumes of cylinders, cones, and spheres.

Tackling Irrational Numbers

In eighth grade, irrational numbers appear. An *irrational number* is a number that you can't write as a fraction with whole numbers in the numerator and the denominator, in contrast with *rational numbers,* which you can (refer to Chapter 12 for more information about rational numbers). The number $\frac{1}{2}$ is rational because it has a whole number in the numerator and in the denominator. The number 3 is rational because you can write it as $\frac{3}{1}$. In eighth grade, students learn that you can't write all numbers this way. The number π is a commonly cited an example of an irrational number.

Eighth graders just need to come to terms with the fact that irrational numbers exist. Knowing that the famous irrational numbers are in fact irrational is also good enough. Indeed, a demonstration that π is irrational requires college-level mathematics. Demonstrating that the square root of 2 is irrational is accessible to high school students.

While students are considering irrational numbers in eighth grade, they're also developing techniques for converting repeating decimals into their fraction forms. A simple technique for this is to notice the following decimal forms of fractions:

- $\frac{1}{9} = 0.11111\ldots$
- $\frac{1}{99} = 0.010101\ldots$
- $\frac{1}{999} = 0.001001\ldots$

These numbers have a pattern. The number of digits in the denominator corresponds to the size of the repeating part of the decimal (called the *repetend*). A decimal such as $0.345345\ldots$ has a three-digit repetend, so $0.345345\ldots = \frac{345}{999}$. You may then simplify the resulting fraction if you like.

Using Exponents Equations

Eighth grade is the time when the shift from arithmetic to algebra is completed. The study of the number system is mostly finished in this year by introducing irrational numbers (refer to the previous section), which leaves room for digging into algebra.

The focus of algebra in eighth grade is different in the Common Core than you may remember from your own ninth-grade Algebra I course. A traditional ninth-grade algebra course includes techniques for dealing with quadratic functions, for example, but this isn't the case in the eighth-grade standards. Instead, the focus is on linear relationships and on the meaning of the symbols and techniques that develop in high school algebra.

These sections give you an overview of eighth-grade algebra by discussing linear relationships, systems of linear equations, the meaning of exponents, and rules for working with exponents.

Looking at linear relationships

The algebra of eighth grade is mostly linear. *Linear relationships,* which are equations with straight-line graphs, are closely related to the proportional relationships that students worked with in seventh grade (see Chapter 12). A proportional relationship has a graph that makes a straight line *and* goes through the point (0,0). The graph of a linear relationship *can* go through the point (0,0) but doesn't have to. A linear relationship does have to have a constant rate of change between the two variables.

When eighth graders study linear relationships, they build relationships among all of the following representations — equations, graphs, tables, and contexts. Students look for the big ideas — such as slope and rate — in each of these representations in order to build a better understanding of the overall idea: linear relationships.

Considering linear equations in all their forms

Linear relationships can appear in any of the following forms. You can transform any one of these forms into the others, and different forms have different purposes:

- ✔ **Slope-intercept form:** The *slope-intercept form,* represented as $y = mx + b$, is useful for describing situations involving change or movement.

  For example, if your little brother runs 10 feet per second and you give him an 80-foot head start in a race, then you can describe his distance from the starting line (y) as $y = 10x + 80$. Your little brother's head start appears as the value of y when x is zero. His running speed appears as the coefficient of x. (A *coefficient* is a number that is multiplied by a variable; $10x$ means *ten times x,* so you can say that 10 is the *coefficient* of x.)

- ✔ **Standard form:** *Standard form,* represented as $ax + by = c$, is the more natural form for situations where the two variables are being added together.

For example, if you and I eat lunch at a restaurant together, and our total bill is $33.18, we can write $x + y = 33.18$, where x represents my part of the bill and y represents your part of the bill. This equation represents all possible combinations of our lunch costs for this particular total. If you add a constraint, such as that my lunch cost twice as much as yours — written as $x = 2y$ or $x - 2y = 0$ — then the two equations together form a system with (in this case) a single solution.

✔ **Point-slope form:** *Point-slope form,* represented as $(y - y_1) = m(x - x_1)$, is a lot like slope-intercept form, except that it assumes you know a point other than the starting point.

Say that you knew that your brother was 90 feet from the starting line after 1 second, but you didn't know how big his head start was. Then you could write $(y - 90) = 10\ (x - 1)$. After a little of algebraic manipulation:

$$(y - 90) = 10(x - 1)$$
$$y - 90 = 10x - 10$$
$$y = 10x - 10 + 90$$
$$y = 10x + 80$$

This equation comes out the same as the one in point-slope form: $y = 10x + 80$. The two forms are equivalent.

Linear equations come in several forms, and these forms have different uses. But the hallmark of a linear equation (or linear function; refer to the "Studying Functions for the First Time" section) is that there is a constant rate that relates the two variables — that is, when the value of one variable changes by a certain amount, the other variable changes by a certain amount, and the ratio of these amounts is unchanging. In the younger-brother-running-race example, the rate is *feet per second*. In the first restaurant bill example, $x + y = 33.18$, the rate is disguised but still there. If our bills add to a total of $33.18, then every dollar more that I spend on lunch is one dollar less that you spend. So there is a rate of –1 dollars on your bill per dollar on my bill.

Eyeing linear graphs and tables

Equations aren't the only representations that students use as they study linear relationships. They also make graphs and tables. Figure 13-1 shows a graph and a table for the running race equation. A linear relationship always appears as a straight line on the graph.

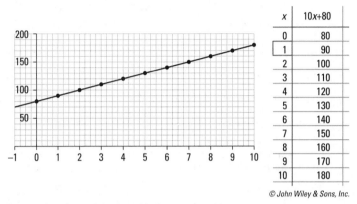

x	10x+80
0	80
1	90
2	100
3	110
4	120
5	130
6	140
7	150
8	160
9	170
10	180

© John Wiley & Sons, Inc.

Figure 13-1: A graph and a table for $y = 10x + 80$.

When students look at graphs, they notice an important idea: The *slope* of a line is the same as the *rate*. Too often in algebra, students come to understand slope in a limited way — as a measure of the *steepness* of a line. Although this is a useful thing to know, slope is much more deeply connected to rates than this idea suggests.

Considering contexts

You can always compute the slope of a line using the classic phrasing *rise over run*. That is: slope $= \frac{rise}{run} = \frac{\Delta y}{\Delta x}$, where Δy stands for the change in the *y*-values, and Δx represents the corresponding change in *x*-values. Computing slope by using rise over run is exactly like the computation you would do if you were figuring out your walking rate. You divide the distance you walked (the Δy) and by the time that elapsed (the Δx).

Similarly, the rate (or slope) describes patterns in the table. In Figure 13-1, the *x*-values count by 1 as you read down. It doesn't have to be the case that a table's *x*-values count by 1, but it's a useful and common way of making an algebraic table. The differences in the consecutive *y*-values are all 10. So for adjacent lines in the table, $\Delta y = 10$, and $\Delta x = 1$. That means $\frac{\Delta y}{\Delta x} = \frac{10}{1} = 10$, which is the same as the slope of the line.

I got the power! Using exponents

Addition and subtraction are operations — ways of using two numbers to compute a third. Multiplication and division are also operations. In algebra, a fifth operation is important — *exponentiation*, which you can think of as repeated multiplication. To help you understand the significance of this, allow me to give you an analogy.

You know that $4 \cdot 3 = 12$, because there are 12 things in 4 groups of 3. If you didn't *know* the product $4 \cdot 3$, you can find it in several ways. You could lay out 4 groups of 3 things and count them one by one, for example. Or you could use the *associative property* of multiplication (which means that to find $a \cdot b \cdot c$, you can either multiply a and b first, or you can multiply b and c first — the final product is the same either way; check out Chapter 8 for more information) to think of $4 \cdot 3$ as $2 \cdot (2 \cdot 3)$, so $4 \cdot 3$ is twice as much $2 \cdot 3$. Finally, you could think $4 \cdot 3 = 3 + 3 + 3 + 3$. That is, one way to compute products is by using repeated addition.

It's the same with exponents: $4^3 = 64$. From 4 and 3, you compute a third number, 64. Just as you *can* compute $4 \cdot 3$ by using repeated addition, you can compute 4^3 using repeated multiplication: $4^3 = 4 \cdot 4 \cdot 4$. But there are other ways too, and these ways depend on properties of exponentiation as an operation. You can double $2 \cdot 3$ to get $4 \cdot 3$ using the associative property of multiplication, and properties of exponentiation allow you to relate 4^3 to 2^3. These properties are known as *rules for operating with exponents*.

Three major rules appear in eighth grade. In the following statements, A is presumed to be a positive number:

- $A^B \cdot A^C = A^{B+C}$
- $A^{BC} = \left(A^B\right)^C$
- $A^{-B} = \frac{1}{A^B}$

You can understand these rules better by way of examples. You can see the first rule, that $A^B \cdot A^C = A^{B+C}$, by thinking of $3^2 \cdot 3^4 = (3 \cdot 3) \cdot (3 \cdot 3 \cdot 3 \cdot 3) = 3^6$. Six threes are multiplied. The second rule you can see by thinking about $(3^2) = 3^2 \cdot 3^2 \cdot 3^2 \cdot 3^2 = 3^8$. Eight threes are multiplied together. The third rule is the logical consequence of the first rule, and of the fact that $A^0 = 1$

when A is any positive number. Here's why: $3^2 \cdot 3^{-2} = 3^0$ by the first rule. Then $3^0 = 1$, so 3^{-2} has to be the reciprocal of 3^2.

Each of these rules is useful going in both directions. You don't have to view these equations as machines that transform the left-hand side into the right-hand side. Instead each side of each equation has the same value as the other side. Sometimes you have something that looks like this: $A^B \cdot A^C$, and it's useful to write it as A^{B+C}. Sometimes it goes the other way around. What matters is the equivalence — or *sameness* — of both sides of each equation.

Eighth graders use exponents to express very large and very small numbers, especially using scientific notation. *Scientific notation* emphasizes the size of a number by writing it using a number greater than 1 but less than 10, a fixed number of decimal places, and a power of 10.

For example, you write 3,456 as 3.456×10^3 in scientific notation. Doing so is especially useful for working with very large numbers (such as 1.52×10^{11}, which is approximately the number of meters you are from the sun right now), and with very small numbers (such as 1.2×10^{-8}, which is approximately the diameter of a virus in meters). Scientific notation makes it much easier to pay attention to the size of these numbers than writing out all those zeroes. When students compute with numbers in scientific notation, the rules for operating with exponents simplify these computations greatly.

Solving two equations at once

Eighth graders write and solve systems of linear equations. They use symbolic techniques and read approximate solutions from graphs. A *system of equations* in eighth grade is a set of two equations, each using the same two variables, and the constraint that the same values for these variables must solve both equations.

For example, my local butcher shop sells 90 percent lean ground beef for $3.89 per pound and 85 percent lean ground beef for $3.79 per pound. If you let x be the price per pound of purely lean beef and y be the price per pound of pure beef fat, then you can write the following system of linear equations (the curly bracket on the left indicates that these two equations go together and that you're interested in any x and y values that make *both* equations true at the same time):

$$\begin{cases} .9x + .1y = 3.89 \\ .85x + .15y = 3.69 \end{cases}$$

Writing a system of equations such as this one requires making a bunch of assumptions about the real world. For example, you have to assume that there is such a thing as purely lean beef and that the butcher store would sell it to you. You also have to assume that ground beef is made by mixing purely lean beef together with pure beef fat and so on. Lots of assumptions — but even if these assumptions aren't all valid, the mathematical solution to the system still gives a starting point for talking about how these pricing decisions get made.

In this case, you can read an approximate solution off the graph in Figure 13-2 by noticing where the two lines cross each other. Purely lean beef should be priced at $4.29 per pound and pure beef fat at $0.29 per pound.

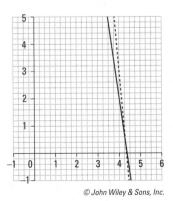

© John Wiley & Sons, Inc.

Figure 13-2: A graphical solution to a system of equations.

I briefly describe a symbolic technique for solving this system of equations. Imagine you bought 1.5 pounds of 90 percent lean ground beef. Then the first equation in the system changes to show that you have 1.35 pounds of purely lean beef, 0.15 pounds of pure beef fat, and that you would pay $5.835. The new system of equations looks like this:

$$\begin{cases} 1.35x + .15y = 5.835 \\ .85x + .15y = 3.69 \end{cases}$$

Both of these equations involve 0.15 pounds of pure beef fat, which means that the additional cost of the 90 percent lean ground beef (the 5.835 in the top equation) comes only from the additional purely lean beef. This additional purely lean beef is half a pound (the difference between 0.85 and 1.35 pounds). You can write this new relationship as $0.5x = 2.145$ (the additional half pound of purely lean beef costs $2.145). You can solve for x to get $x = 4.29$. This matches the value you can see on the graph. This kind of thinking develops into the algebraic technique of *elimination*. Eighth graders solve systems of linear equations by elimination as well as by working with graphs.

Delving into Functions

Many mathematicians would agree that the most important objects in mathematics aren't numbers; they're functions. Mathematicians use the idea of a *function* to describe operations such as addition and multiplication, transformations of geometric figures, relationships between variables, and many other things. Eighth grade is the first time students meet the term *function*. The following sections examine what eighth graders do with functions.

Defining functions

A *function* is a rule for pairing things up with each other. A function has inputs, it has outputs, and it pairs the inputs with the outputs. There is one important restriction to this pairing: Each input can be paired with only one output.

An example of something that isn't a function is $y = \pm\sqrt{x}$. In this case, the $\pm$ symbol means plus or minus and allows for more than one y-value for the same x-value. This new rule generates these ordered pairs, for example: (4, 2) and (4, –2). The same input (4) is paired with two different outputs (2 and –2), which means that $y = \pm\sqrt{x}$ isn't a function.

For example, the function $y = x^2$ pairs numbers with each other. The x-values are the inputs and the y-values are the outputs. Often, people write the pairs using this notation: (x, y), which means that (1, 1), (2, 4), and (–3, 9) are all pairs generated by the rule $y = x^2$. As required, this function has only one output for each input. There is no requirement the other way around, though. It's okay that (2, 4) and (–2, 4) are both pairs generated by this rule. In this case, one *output* is paired with two different *inputs*, which can happen (and it often does) with a function.

Students learn to identify functions and non-functions from their graphs. If two points are on a graph so that one point is directly above the other, it means that the same x-value is paired with two different y-values, so the graph doesn't describe a function. Sometimes, it's referred to as the *vertical line test*. The graph on the left in Figure 13-3 is of $y = x^2$ and the graph on the right is of $y = \pm\sqrt{x}$.

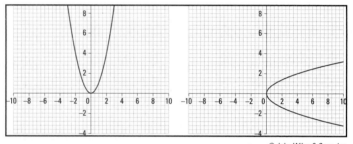

© John Wiley & Sons, Inc.

Figure 13-3: On the left, a function. On the right, not a function.

Knowing the definition of a function is a good thing. In mathematics, definitions help you to sort the world out in a matter of fact way. Something either is a function or it isn't a function because it either fits the definition or it doesn't.

Most people, however, don't go through life — or even through math class — by constantly thinking about definitions. Getting better at referring to and using definitions is a Standard for Mathematical Practice (refer to Chapter 3 for more on these standards), but most people refer first to a set of mental images when they think about categorizing things. People tend to ask whether something looks like a function or not before they refer to the definition. Because students spend so much time with functions such as the one on the left in Figure 13-3, they tend to miss the strange cases of functions such as those in Figure 13-4.

Each of the graphs in Figure 13-4 represents a function because it fits the definition of pairing x-values with y-values in such a way that each x-value gets only one y-value. However, none of the graphs in Figure 13-4 looks like the functions that students encounter as they go about their average day in a traditional algebra class. In a Common Core classroom, students encounter (and even produce) these kinds of examples in the service of better understanding functions.

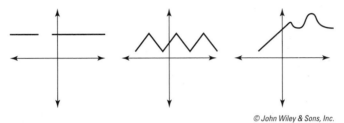

© John Wiley & Sons, Inc.

Figure 13-4: Examples of functions.

Figuring out a function's function

An important way to think about functions is as relationships between variables. If you think of a function as a relationship, you can keep an eye out for useful features. These features include whether (and where) a function is increasing or decreasing, whether a function is linear or not (refer to the earlier section "Toeing the line on linear relationships"), and so on.

You can read a graph from left to right. If the graph rises as you read from left to right, it means that the y-values are *increasing*. If the graph falls as you read from left to right, the y-values are *decreasing*. If the graph is horizontal, the y-values are *constant*. A function can be increasing for all x-values (such as the line $y = 2x + 2$) or it can be increasing in some places and decreasing in other places (such as $y = x^2 + 2$) Refer to Figure 13-5 as an example.

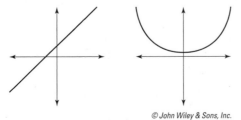

© John Wiley & Sons, Inc.

Figure 13-5: Graph (a) is increasing for all x-values; graph (b) isn't.

In eighth grade, students may simply describe the change they see in graphs of functions, and often these functions don't have symbolic form.

Doing the Ancient Greeks Proud

School geometry is directly descended from the work of the ancient Greeks; in eighth grade, it's especially true of Euclid and Pythagoras (and his colleagues). Euclid wrote *The Elements*, which is considered a foundational text for both geometry and proof in math. The work of Pythagoras remains influential because of the theorem that bears his name. These sections delve deeper into these areas of geometry for eighth graders.

Deciding what's the same

In eighth grade, students study congruence and similarity as two ways of talking about how two shapes are the same. In math, there are many kinds of sameness. In geometry, two major kinds of sameness are congruence and similarity.

Two shapes are *congruent* if you can move one so that it perfectly matches the other one without stretching or deforming it. Two shapes are *similar* if you can do the same thing, except that you're allowed to stretch or shrink the shapes proportionally — you can double *all* the lengths on one of the shapes, for example.

Eighth graders need to carefully specify the size of the rotation (measured in degrees or in fractions of a full turn) and the distance of the slide (measured in any linear units, such as inches or centimeters) necessary for moving one shape so that it matches up with a congruent one. They also need to identify precisely the factor by which a shape needs to be stretched or shrunk to match a similar shape.

Look at Figure 13-6 and determine which are the same.

In this figure, only triangles A and B are congruent. But triangle C is similar both to A and to B. So which shapes are the same? It depends on what you mean by *same*. This confusion is why mathematicians use the words *congruent* and *similar* — the meanings of these words are much more precise than the meaning of *same*.

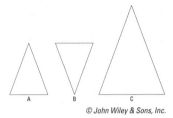

© John Wiley & Sons, Inc.

Figure 13-6: Examine these triangles to see which are the same.

Measuring the hypotenuse

The Pythagorean theorem is a triumph of human ingenuity. Whether you attribute it to Pythagoras, the Babylonians, or to Indian or Chinese scholars, the idea is profound. You can calculate a length without directly measuring it. The Pythagorean theorem is a special case of indirect measurement, but it turns out to be astonishingly useful.

Eighth grade is when students learn the Pythagorean theorem in the Common Core State Standards. The *Pythagorean theorem* is this: In a right triangle, the sum of the squares of the lengths of the two legs is equal to the square of the length of the hypotenuse. This sentence is complicated, but the idea isn't nearly so complicated. Figure 13-7 is simpler to interpret.

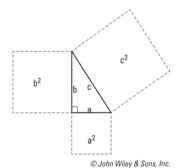

© John Wiley & Sons, Inc.

Figure 13-7: An example of the Pythagorean theorem.

The area of each smaller square is equal to the square of the length of a leg, where a *leg* is one of the two shorter sides of the right triangle. The area of the largest square is equal to the square of the length of the hypotenuse, where the *hypotenuse* is the length of the longest side of a right triangle. The Pythagorean theorem says that the areas of the two smaller squares add up to be the same as the area of the largest square.

In symbols, this relationship is stated even more compactly. If a and b are lengths of the legs of a right triangle and if c is the length of the hypotenuse, then $a^2 + b^2 = c^2$.

Two of the most common errors students make as they become accustomed to using the Pythagorean theorem are thinking that $a + b = c$, and forgetting that the sum of a^2 and b^2 is the *square* of the length c. You can lessen the likelihood of each of these errors by sketching a diagram such as the one in Figure 13-7, with numbers substituted for the numbers in the problem you're working. This diagram helps focus your attention on the relationship between the side lengths of the triangle and the areas of the squares.

Eighth graders learn a proof of the Pythagorean theorem. Figure 13-8 shows two big squares with the same area. Each of these big squares is subdivided into four congruent right triangles and some other stuff. In the big square on the left, things are arranged so that the other stuff consists of a square on each leg of the right triangle. In the big square on the right, the other stuff consists of a square on the hypotenuse. The conclusion is that the two squares on the legs have the same combined area as the square on the hypotenuse, so $a^2 + b^2 = c^2$. The key to this proof is that the only thing special about the triangles is that they're right triangles.

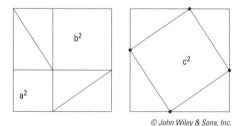

© John Wiley & Sons, Inc.

Figure 13-8: A proof of the Pythagorean theorem.

The converse of a theorem in math isn't generally true, but the converse of the Pythagorean theorem is true. The Pythagorean theorem starts with a right triangle and concludes that the side lengths have the relationship $a^2 + b^2 = c^2$. The *converse* starts with a triangle whose sides have the relationship $a^2 + b^2 = c^2$ and concludes that the triangle is right.

One final word about the Pythagorean theorem: The relationship $a^2 + b^2 = c^2$ is *not* true for triangles that aren't right. Instead, if the largest angle in a triangle is acute, then $a^2 + b^2 > c^2$, and if it's an obtuse angle, then $a^2 + b^2 < c^2$ (refer to Figure 13-9 for an example).

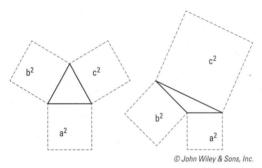

Figure 13-9: $a^2 + b^2$ doesn't equal c^2 in these triangles.

Finding tricky volumes

In eighth grade, students find the volumes of cylinders, cones, and spheres. As with the development of area relationships in earlier grades, the development of volume relationships also depends on simpler relationships. In this case, the formula for volume of a cylinder depends on the volume of a prism (refer to Chapter 11).

You can find the volume of a prism by finding the area of the base and then multiplying by the height. You can think of the area of the base as the number of unit cubes that would fit in one layer on the bottom of the prism and the height as the number of layers that fill it. When the prism is rectangular, then the area of the base is found by multiplying the length and width of the prism, so $V = l \cdot w \cdot h$. This same principle applies to cylinders — the area of the base uses the formula for area of a circle (check out Chapter 12), so $V = \pi r^2 h$. Figure 13-10 shows how a cylinder, like a prism, consists of layers.

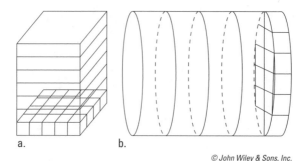

Figure 13-10: A prism and a cylinder — both are composed of layers.

Volume formulas for cones and for spheres come from the volume of a cylinder. A cone has one-third the volume of a cylinder with the same radius and height as the cone, so $V = \frac{1}{3}\pi r^2 h$ for a cone. A sphere has two-thirds the volume of a cylinder with the same radius and height as the sphere. The subtle difference for a sphere is that its height is twice its radius, so $V = \frac{2}{3}\pi r^2 2r$, or more simply $V = \frac{4}{3}\pi r^3$.

Addressing Bivariate Data

Statistics in eighth grade connects the work of the previous grades with the linear relationships work of eighth-grade algebra. Students collect *bivariate data* in order to look for relationships between two variables, then they fit linear models to this data and ask about the quality of the fit.

For example, the relationship between height and age in people is *bivariate*, meaning that it has two variables. Fitting a linear relationship to this data suggests a belief that people get taller as they get older, and that this rate of growth is fairly constant over time. Eighth graders collect data for this sort of question, make estimates about good linear models, and debate the merits of these models. They question assumptions and come to understand that statistical models may be good for making ballpark estimate predictions but shouldn't be expected to produce precise predictions. Also they conclude that all models have limits. Maybe linear growth of height as a function of age is sensible for teenagers, but it certainly doesn't describe older adults.

Chapter 14

Looking at Advanced Math: High School and Beyond

. . . .

In This Chapter

▶ Understanding the purpose of high school math

▶ Structuring high school math programs

▶ Sampling the big topics in high school math

. . . .

*T*his chapter addresses the structure and main ideas of high school math, including the purpose of studying math in high school. (***Hint:*** It's about preparing for life beyond the high school classroom.) In this chapter, you read about the main subject areas in high school, as well as the various ways these subjects relate to each other, and how they can get built into three- or four-year high school programs. Because this one chapter deals with three or four years worth of math, I don't cover the topics in detail like I do in the other chapters. Nevertheless, you should find enough specificity to begin to understand how the K–8 program develops into high school math and how high school math can be useful for students as they head into careers and college.

Knowing What College Ready Means

The major stated goal of the Common Core State Standards is to ensure that all American high school graduates are *college and career ready.* Of course, students need many more things in order to be college and career ready than they will learn in their math classes. (English and math are the only subjects in the Common Core, but a certain amount of science seems important for nearly all careers, for example.)

College and career ready is a terrific reminder to teachers and to school, district, and state-level decision makers that school has a purpose beyond making kids good at school. School is supposed to prepare students for their lives beyond graduation. The phrase *college and career ready* can help everyone maintain a focus on what's important and useful after the diploma has been earned.

Being *college ready* in mathematics probably means different things to different people. At a minimum, it should include these things:

✔ Being able to do high school (and middle and elementary school) math outside of the math classroom

✔ Being prepared for success in college-level math and science courses

Notice that this short list doesn't revolve around (or even include) earning credits in algebra, geometry, and so on. The system of earning credits isn't going away from American high schools and colleges anytime soon, of course. But the focus on college and career preparedness is intended to remind everyone involved with schools of the reasons why students take these courses. Students shouldn't enroll in a course for the sole purpose of earning a credit of some kind. They should enroll in a course in order to learn something that will be useful to them immediately or later in life.

So college and career readiness is not about credits; it's about learning instead. That's the good news about the phrase: *college and career ready*. Here's the bad news: There is no general agreement about what exactly a *college-level* course is. The following sections examine some of the perspectives and the practical meanings of this lack of consensus.

Examining some truths about college math placement

Before many students start college classes, they often take a math placement test without specific preparation. Many perform more poorly than if they would have studied. As a result, they're often placed in a lower math course than is appropriate, and they end up wasting a semester of time and tuition.

In the worst-case scenario, the student places into a developmental math course, where *developmental* is the polite term for *remedial*. Now, some students are well served in these courses, but as a group, developmental math courses have very low pass rates and are frequently obstacles to students moving on in their college work and eventually to earning degrees.

Math placement at the college level has three main systems:

- ✔ **High school GPA:** The best predictor of success in college-level math courses is high school GPA (but not by a lot).

- ✔ **SAT or ACT scores:** These tests that most college-bound students take as high school students are pretty good indicators of which college math courses are appropriate for beginning college students.

- ✔ **Placement tests:** Colleges frequently administer these as a routine part of orientation programming. A few hours reviewing high school algebra a few days or weeks before taking a placement test can make a huge difference. Many resources are available as books or online for this sort of review.

Additionally, prior coursework such as an Advanced Placement (AP) class or credits from a program that offers college-level courses to high school students can impact course placement for students entering college.

The fact that placement tests aren't strong indicators of success, together with their widespread use, is partly a reflection of a lack of consensus about the meaning of *college-level math*. At some colleges, *college-level* means calculus; at others it means math for liberal arts or college algebra, which is a lot like high school Algebra 2.

College readiness — from the college's perspective — is unfortunately a relative concept rather than an objective measurable reality.

Integrating or not

In the late 1990s, there was a big hullabaloo about the idea of an integrated math program at the high school level. An *integrated math program* involves studying appropriate topics at each grade level rather than segregating algebra into ninth and eleventh grades, geometry into tenth grade, and so on. It may mean studying parabolas as geometry objects, not just algebraic ones, in ninth grade.

An integrated math program may look quite different from a traditional one. Perhaps the chief difference from a parent's perspective is in the names of the courses. Students may enroll in Math 9 instead of Algebra, Math 10 instead of Geometry, and so on. More fundamental differences include helping students notice similarities and common threads among math subjects, where traditionally these courses are seen as separate and distinct. A big concern about integrated math programs has been whether colleges would accept a course titled something like Math 9 for admissions.

I have taught at Big Ten universities, a smaller state college, and now a community college. I can assure you that these issues have been worked out. Students taking courses with titles such as Math 9 and Math 10 aren't at a disadvantage relative to students with Algebra and Geometry on their high school transcripts. In fact, the standards movement (of which Common Core is a part) has been instrumental in alleviating these concerns because high school math is now much more standardized for high school graduates than it was 20 years ago. Whatever the course names, the standards are the same.

Looking at integrated math

An *integrated math program* focuses on studying topics at each grade level rather than segregating algebra into ninth and eleventh grades, geometry into tenth grade, and so on. This type of program may have students consider a question such as, "Are all parabolas similar?" A parabola is a curve in the plane defined by a quadratic equation of the form, $y = ax^2 + bx + c$. Similar is a geometry idea (refer to Chapter 13 for more details); it refers to shapes that have proportional measures. All squares are similar because the ratio of two sides of a square (any square) is 1:1. A 3-by-8 rectangle is similar to a 6-by-16 rectangle because they both have a 3:8 side ratio. Similar is a bit more abstract when you're describing infinite objects such as parabolas, so this type of question requires students to learn and to apply a number of challenging algebra and geometry skills.

In addition to $y = ax^2 + bx + c$, a *parabola* can also be defined by $x = ay^2 + by + c$, and by versions of these two equations that rotate the parabola to face in any direction. The standard form $y = ax^2 + bx + c$ gives a parabola and represents y as a function of x, and so is useful in algebra, where functions are major objects of study (check out the section "Studying function families" later in this chapter).

Another result of an integrated math program may be that students study *proofs* in mathematics outside of geometry class. In a traditionally structured math program, one of the major goals of tenth grade geometry is proof writing. An integrated program is more likely to ask students to prove things at all grade levels. A proof is a form of a mathematical argument, and "constructing viable arguments" is a Standard for Mathematical Practice (as I explain in Chapter 3).

The Common Core State Standards don't take a stand on the question of whether to integrate math programs. There are geometry standards, algebra standards, statistics standards, and so on. But there is no requirement that each set of standards makes up a course, nor that a course must draw from across these sets. An appendix to the high school standards helps school and district directors make these kinds of judgments, but the standards themselves don't dictate whether to implement integrated math programs.

Marching to calculus

The traditional organization of high school math courses is based on an assumption that preparation for calculus courses is an important goal. In fact, middle school and high school curriculum committees commonly say that one topic or another needs to stay in the curriculum because "they'll need it for calculus" (where *they* refers to students).

Shifting the focus of high school math curriculum decisions from *they'll need it for calculus* to *college and career ready* allows for more useful decision making in schools and districts. Also it allowed the writers of the Common Core State Standards to think beyond this one track that is important for science, engineering, and math majors, but probably not so important for a large part of the high school student population.

Students going straight into the workforce, those transitioning to career training at (for example) a two-year technical college, or those planning to study

(continued)

(continued)

humanities at a four-year college may be better served by a solid preparation in statistics rather than in calculus. The Common Core State Standards focus on college and career readiness values this preparation together with preparing students for calculus when appropriate.

The other big question around calculus is whether — for students who *do* take this course — it can, or should, be offered at the high school level. Whether it *should* is a highly debatable and contentious issue. College math professors commonly notice that they have a large proportion of students in their Calculus I course who are taking it for the second time, having taken it once in high school. At the same time, students can and do learn useful math in a high school calculus course, and they can save a lot of money by taking it in high school instead of in college. However, taking calculus in high school doesn't guarantee that a college will grant credit for the course. A student who takes calculus in high school may still need to take a placement test or take the college's calculus course.

Schools and districts wishing to prepare their students to take calculus when they head to college (which is the most common time for students to take a calculus course — in their first year of college) typically offer a one-semester or even one-year precalculus course to seniors that covers the necessary math for calculus that isn't in the Common Core State Standards.

Schools and districts wishing to offer calculus to high school seniors find creative ways to ensure that ninth, tenth, and eleventh graders get the mathematics they need earlier than average and offer a calculus course in the building, online, or in partnership with a local college. Actually, the truth is that it doesn't require a lot of creativity at all — just including a few extra topics per year in ninth through eleventh grades, and statistics as a second math course in eleventh or twelfth grade for students on this calculus track. Alternatively, some districts offer Algebra 1 to high-achieving eighth graders as yet another way to position students for calculus in high school.

Advancing an Algebraic Agenda

Algebra in the Common Core State Standards doesn't refer only to a single course. Algebra in high school spans across at least two different years — likely three or four for most students. The following sections provide an overview to the important concepts in high school algebra.

Students study algebra in high school as the system of using symbols and other representations to express and to understand relationships between variables. These variables may be abstract, as when students derive the quadratic formula. These variables may be applied, as when students use information about driving speeds and cell phone use to determine how much roadway goes by while you read and reply to a text message.

This perspective on algebra — that it consists of ways of representing relationships — is in contrast to many people's experiences in the past, where algebra may have felt like a set of arbitrary rules for pushing symbols around. In a Common Core classroom, asking about and understanding these relationships will take center stage. The skills involved in moving symbols around remain important, but they may not be the primary objects of study.

Studying function families

A function is an important mathematical object. At an abstract level, a *function* is a matching of things (usually numbers, but it doesn't have to be). The function $y = x^2$ matches the number 2 to the number 4, for example — when $x = 2$, $y = 4$. In order to be a proper *function*, this matching has to be done so that each x-value is matched to only one y-value, although two x-values may both be matched to the same y-value. All of this is terribly abstract, and this book's purpose isn't to make the nuance of functions clear. (For more information on functions, check out *Algebra 2 For Dummies* by Mary Jane Sterling [John Wiley & Sons, Inc.].) Instead, let me give you the takeaway message: one way to think about functions is as a *matching* (or *correspondence*) of individual numbers.

Another way to think about functions is as *relationships* between variables. As x increases, y increases, for example. Or doubling x also doubles y is another example of a relationship between variables. Every function written as $y =$ some expression or other describes a relationship.

These relationships fall into a number of important and useful categories. In high school algebra, these categories and their characteristics are important objects of study. A few of these categories are

✔ Linear functions

✔ Exponential functions

✔ Quadratic functions and other polynomials

✔ Rational functions

Each can be considered a *function family*. When students develop familiarity with these families and their properties, they can use various types of functions to describe some commonly appearing applied mathematical situations.

Certain kinds of mathematicians don't like the *relationship* idea of function because, formally, a function doesn't need to have any kind of relationship between the two variables. In fact, some functions have no predictable relationship at all. The function that matches the roll of a six-sided die with its outcome is an example — the first roll may give a 3, a second roll may give a 2, and so on. You don't have a way to know what the next roll will be — so there is no relationship — but the matching is done in a way that fits the definition of function.

In addition to this collection of basic relationships, students study the reverse of these relationships. If *y* quadruples when *x* doubles, what happens to *x* when *y* doubles? That sort of question involves the idea of *inverse*. Studying inverses introduces whole new families of functions, such as logarithmic functions and roots.

Solving systems

A *system of equations* or *system of inequalities* is a collection that uses the same variables, with the stipulation that the solutions to one equation (or inequality) need to be solutions for the others as well.

An example of a problem that a system of equations can be used to solve involves chocolate sandwich cookies, some regular and some double stuffed. Three of the regular cookies have 160 calories, while two of the doubly stuffed cookies have 140 calories. You can use the following system of equations to represent the caloric relationships:

$$\begin{cases} 6x + 3y = 160 \\ 4x + 4y = 140 \end{cases}$$

In this system of equations, x represents the number of calories in one chocolate wafer and y represents the number of calories in one layer of gooey stuff. The same variables (x and y by name, but *number of calories in a wafer* and *number of calories in a layer of stuff* by meaning) are used in both equations. The assumption that solutions for one equation need to be solutions for the other as well comes from assuming these two things:

 ✔ The wafers in both the regular and doubly stuffed cookies have the same number of calories.

 ✔ The double ones really do have twice as much gooey stuff inside.

If either of these things isn't true, then the system of equations is flawed and the solution to that system won't give the real number of calories in a wafer and in a layer of gooey stuff.

As students progress through high school, they develop increasingly sophisticated strategies for solving systems of equations and of inequalities, and they solve increasingly sophisticated systems.

For example, students solve systems of two linear equations, and then of three linear equations in three variables. Adding one variable increases the level of difficulty substantially, but the basic ideas remain the same. Later, they solve systems that involve a linear equation and a quadratic equation. Doing so requires new techniques. Students also solve systems of two linear inequalities. These last systems have solutions that look quite different from solutions to systems of equations, because these new solutions typically consist of infinitely many points.

Techniques that students use to solve systems of equations include solving them graphically — by finding the point of intersection of two lines. They solve systems of equations symbolically using methods commonly referred to as *substitution* and *elimination*. Matrices are used to solve challenging systems that could involve more than three equations and variables. In these last cases, students typically use technology to find the numerical values of the solution because these computations are quite tedious. The useful thing about these kinds of problems is understanding their structure rather than grinding out the solution.

Using algebraic structure

Algebra can help to reveal how things are built — what different scenarios have in common with each other and what the important differences are. The Common Core State Standards refer to this idea as *algebraic structure*.

When you double the side lengths of a rectangle (or any geometric figure), but leave the angle measures the same, the area grows by a factor of 4. Triple the lengths and the area grows by a factor of 9. Doing so makes the relationship between length and area *quadratic*. Quadratic structure appears in many other guises — as the relationship between the number of vertices of a regular polygon and the number of its diagonals, for example, and as the result of multiplying two linear functions together. When you pay attention to the similarities of these various contexts, you notice that they have the same structure (refer to the earlier section "Studying function families").

Noticing and using algebraic structure requires stepping back from your work and thinking about the big picture. In fact, doing this is an SMP (I discuss these standards in Chapter 3). Rather than looking at a chapter title in their textbook to decide what technique to use, students need to select techniques and ideas that make sense based on the relationships in the problem being considered.

Algebraic structure supports students working with *mathematical modeling*, which is an important part of the Common Core high school curriculum (and across all the grades). When students are building mathematical models, they're typically trying to understand something better that comes from outside the realm of mathematics. They're trying to answer a question that someone might reasonably ask outside of math class, and using mathematics to get a good answer. Part of building a good mathematical model involves seeing the structure of the situation. A student may ask questions such as "How are my variables related?" and "What other problems have I solved that involved similar relationships?" These questions draw the student back from the particulars of the context and help the student to focus on its mathematical structure.

Getting Formal with Geometry

The traditional view of high school geometry is that it is about proof. The basic structure of a standard high school geometry course is inherited from Euclid's *Elements*, a text more than 2,000 years old.

Euclidean geometry and proof remain central parts of geometry in the Common Core State Standards as well, but some more modern ideas are embedded. In particular, *transformational geometry* (where students study what happens when they move shapes and change their size) is more important than it may have been in the past, with students studying the mathematics that results from sliding, flipping, rotating, stretching, and shrinking shapes.

Analytic geometry (where students study shapes on a coordinate grid) also gets more emphasis as students describe shapes using algebraic equations and coordinates.

These sections describe the major topics of high school geometry, including proof, similarity and congruence, and trigonometry.

Proving stuff

The experience many people describe with high school geometry is that it was very different from what came earlier. Whereas algebra may have seemed like an endless collection of steps, geometry was about *proofs* (which are how mathematicians justify what they know to be true). In proof writing, you lack a list of steps to follow, unlike when solving equations in algebra. For some people, this difference meant that geometry felt like a breath of fresh air while for others it felt frightening or confusing. But the difference was palpable for many people.

The Common Core State Standards reduce the size of this difference in two ways:

✔ By introducing more argumentation and proof into earlier grades

✔ By developing the study of geometry more carefully in earlier grades

Proof is a form of mathematical argument. The SMP include "Construct viable arguments and critique the reasoning of others." Because the SMP are for all grade levels, students arrive at the study of high school geometry having had much more experience with mathematical argumentation — and therefore much more experience with proof.

Furthermore, a major gap in standard American curricula for many years has been between *identifying and naming shapes* in the primary grades and *proving theorems* in tenth grade geometry. The Common Core State Standards, by contrast, develop the study of geometry across the elementary and middle grades in ways that deepen the relationships students encounter over time. By the time they reach high school, students in Common Core classrooms have identified and studied relationships among classes of shapes, which leads naturally to studying geometric relationships closely and to proving things in high school geometry.

Using similarity and trigonometry

Students studied similar figures in eighth grade (see Chapter 13). In high school, they use the properties of similarity to study new things, in particular, trigonometry.

The two triangles in Figure 14-1 are *similar*, which means that the measures of corresponding angles in the two triangles are the same, and that their side lengths are proportional. *Proportional* can refer to one of two relationships in this situation:

- ✔ The ratios of corresponding sides between the two triangles are all equivalent, so $\frac{a}{d} = \frac{b}{e} = \frac{c}{f}$. This common ratio is sometimes referred to as a *scale factor*.

- ✔ Each ratio of two sides within a triangle is equivalent to the ratio of the corresponding sides in the other triangle, so $\frac{a}{c} = \frac{d}{f}$ for example.

These relationships hold for similar triangles in general (and for similar polygons of all kinds, actually). The triangles in Figure 14-1 have an additional property: They're right triangles. In this case, the ratios $\frac{a}{c}$ and $\frac{d}{f}$ are fixed, and their value depends on the measure of the angle at A (and at D, which is the same measure because the triangles are similar). This ratio is *sin* (A), which you read as *sine of A*.

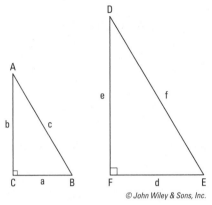

© John Wiley & Sons, Inc.

Figure 14-1: Two similar right triangles.

Because the ratios of sides of right triangles were figured out long ago and because these ratios are extremely useful for measuring things, the relationships between angle measures and the ratios of sides of the corresponding right triangles have been given names. *Sine, cosine,* and *tangent* are the main ratios, but *secant, cosecant,* and *cotangent* are similarly defined.

Trigonometry is the study of these functions. The Common Core State Standards don't have a separate set of trigonometry standards. Instead, students study the trigonometry functions as part of their work in geometry — whether in a dedicated geometry course or in an integrated math program across several grades. Either way, a typical precalculus course extends the study of these ratios to angle measure values that don't make sense in right triangles, such as negative numbers and numbers greater than 90°. The Common Core State Standards don't include precalculus standards, so they contain only the right triangle trigonometry that I describe here.

Describing shapes with algebra

Certain shapes can be represented by a short algebraic equation, as long as these shapes are drawn in the coordinate plane. A line can be described with the equation $y = mx + b$, for example, or a circle with the equation $a = x^2 + y^2$. Having a way to describe shapes with algebra allows students to ask and to answer new questions about the properties of these shapes.

This works the other way around, too. Considering the graphs of algebra equations as geometric objects allows students to study new shapes in geometry that they wouldn't think to look at otherwise. For example, the equation of a circle is $a = x^2 + y^2$, but if the terms are *subtracted* instead of *added* to get $a = x^2 - y^2$, a whole new shape is discovered — a *hyperbola,* as Figure 14-2 shows.

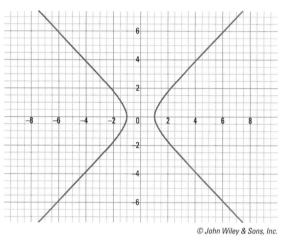

© John Wiley & Sons, Inc.

Figure 14-2: A hyperbola.

Understanding the World, Statistically Speaking

Probability and statistics are forever intertwined. Most interesting statistical questions boil down to a form of the question: "How likely is it that I'm right?" Researchers who have a hunch that a medication or procedure will be effective in treating a disease don't end up with an answer such as "Yes, this always works." Instead, they get a much more nuanced answer that depends on statistics and on probability — something such as "People in this study who got this treatment lived longer on average, and there is a 95 percent chance that this is because of the treatment (but a 5 percent chance that it has nothing to do with the treatment)."

The number of decisions you have to make in your daily life that involve statistical understanding is staggering. Financial and medical decisions in particular rely on statistical reasoning. Unfortunately, the human mind seems to be particularly bad at nuanced statistical reasoning. People tend to prefer absolutes — always do this, never do that, and so on.

Of course, this is an argument in favor of teaching statistical reasoning, not an argument against it. In the Common Core State Standards, students study quite of bit of statistics, whether in a dedicated course or spread across several years in an integrated math program.

Quantifying data in one or two variables

Data measures become important anytime you have multiple measures of the same variable. "How tall am I?" is a measurement question. To answer it, you measure yourself, get a value, and you're done. "How tall are ninth graders?" is a statistical question, not a measurement question. You can still measure as part of answering the question. You can, for example, measure each ninth grader at your school. But all those measurements don't really answer the question.

To answer a question such as this one in a meaningful way, you need to do something with these measurements. As soon as you start doing things with a set of measurements, you're doing *data analysis*. This example has only one variable — height. To answer the question, you could make a graph of some sort (a bar graph, perhaps, or a histogram) representing all of these heights. That might lead you to notice that most ninth graders are between 64 and 69 inches tall.

Alternatively, you may decide to compute a single value to represent all of these measurements. The *mean (average)*, *median* (middle value in an ordered list), and mode (the most common value) each give somewhat different information about a dataset, but each can be used to describe the height of the typical ninth grader. You could compute the *standard deviation* (a measure of spread) in order to describe how much ninth graders' heights vary (answer: a lot more than kindergarteners' heights).

All of these are ways of quantifying data in one variable. In high school, students also quantify relationships between two variables. Continuing with the height theme, a simplistic example could be asking whether taller people have bigger feet. As before, you would need to collect a whole bunch of measurements in order to answer this question, but the measurements themselves don't answer it.

In this case, a better representation than a bar graph would be a *scatter plot* (a graph of points in the coordinate plane) with height on the x-axis and foot length on the y-axis. If you decide a relationship exists and that taller people do have bigger feet, then you'll want to get more specific. You could ask, "How does foot length grow in relation to height?" High schoolers answer this sort of question by fitting models to their data; the most common model is linear. A linear equation such as $y = .3x - 8.9$ would mean that your foot grows .3 inches longer for every 1 inch taller you grow (on average).

Understanding the relationship between data and probability

Statistics depends on probability, and probability depends on statistics. *Probability* is about "How likely?" questions. *Statistics* is about "What is the relationship?" questions about sets of data. The relationships you identify in answering statistics questions always involve probability. And vice versa — answers to probability questions are based on having collected data. Here is an extended example that demonstrates this reciprocal relationship between data and probability.

From time to time, soft drink companies run promotions in which you can win a prize if a specific code is printed inside the can. You can't see the code until you buy and drink the beverage. You know that your can has a chance of winning when you take it from the store shelf, but you don't know whether it's a winning can. A typical claim in this kind of promotion is that "one in six wins," which means that of all the cans in this promotion, one-sixth of them are winners.

That one-sixth of the cans are winners is a statistics claim. Each can is measured (in a sense) as either a winner or a loser. The data on all of the cans is summarized — that's doing statistics. Each can has a one-sixth chance of being a winner — that's a probability claim.

Now you decide to buy a six-pack of this beverage. You may do so believing that you'll be sure to get a winner. The probability question there is: "How likely am I to get a winner if I buy a six-pack?" Your sureness about winning when you buy a six-pack is based on statistics — the data summary that one-sixth of the cans are winners.

Even if you think you'll certainly get a winner, you're answering a probability question; you're saying to yourself that the answer is 100 percent. Now you buy yourself a six-pack and are surprised to find no winner in it. This data should force you to change your thinking about the probability.

In the end, the probability of getting at least one winner when you buy a six-pack is about $\frac{2}{3}$. High school students can find this result using probability theory. $\left(1-\left(\frac{5}{6}\right)^6\right)$ is one computation that works here.) Also they can find it using statistics. Rolling a die once can represent one can, which is either a winner (if you roll a 1) or a loser (if you roll anything else). Rolling a die six times represents buying a six-pack. If students roll a large number of groups of six rolls and if they keep track of the number of six-packs with wins, they'll end up with something quite close to $\frac{2}{3}$ of these six-packs having at least one winner.

In the Common Core State Standards, probability and statistics are intertwined in this way. High school students commit to probability models and then test those models by collecting data. Similarly, they use probability to describe how likely the conclusions they draw from their data are.

Using probability to make decisions

One of the most powerful tools in the probability toolkit is expected value. *Expected value* describes the average payoff for a chance event. Many, many decisions are made based on expected value.

State lotteries and casinos provide the most straightforward cases of expected value. Imagine that you work for a lottery and you have to create a new scratch-off game. At a minimum, you would want to design a game that earns more money from ticket sales than it gives away in prizes. So if the cost of a ticket is a dollar, then you need to design your game so that the total of all of the prizes, averaged over all the tickets, comes out to less than a dollar. That is, the *expected value* of a ticket must be less than a dollar. In reality, it needs to be quite a bit less to pay for the costs of the lottery and to raise money for the state.

With a scratch-off game, things are simple because you can control exactly how many prizes you give out and what types they'll be. A game where people choose numbers and the lottery randomly selects winning numbers makes things more complicated. In that kind of game, the expected value of a ticket doesn't precisely predict the outcomes because the lottery has no control over the numbers people pick or over the winning numbers. Sometimes many people happen to pick the winning numbers, and sometimes few people do, which means that sometimes the average ticket will win more than the expected value, and sometimes it will win less than the expected value.

But even in the drawing-numbers lottery games, things are relatively simple compared to the expected value situations the insurance industry has to wrestle with. Some sample questions that people in the insurance industry need to be able to answer include

- ✔ How likely is another direct hit of a hurricane on New Orleans, and how much damage would it do?

- ✔ How likely is this healthy 44-year-old male to die this year (of all causes combined) and what sum of money would ensure the financial security of his family?

- ✔ How likely is a diabetic 71-year-old woman to need medical care beyond routine check-ups this year, and what will be the cost of that additional care if it is needed?

These questions all involve complicated expected value computations. In each case, there is a probability and a cost. In each case, the expected value is used to decide how much money to charge for insurance — whether property insurance, life insurance, or health insurance.

High school students develop these ways of thinking about probability in general and expected value in particular. Probability isn't just about flipping coins and rolling dice. It's also about trying to determine what is fair in an uncertain world.

Part IV

The Part of Tens

Go to www.dummies.com/extras/
commoncoremathforparents for a list of ten benefits
of the Common Core Standards.

In this part . . .

✔ Find resources for supporting your child's math learning at home — beyond helping with homework — so that math can become an enjoyable and regular part of your family time.

✔ Get in touch with the help *you* need when *you* have a math question so that you know where to find answers.

✔ Figure out how to set a positive tone about math and your child's achievement so that your child understands that she can be successful.

Chapter 15

Ten Awesome Resources for Parents

In This Chapter

▶ Discovering helpful math resources on the Internet

▶ Finding ways to support your child at home

▶ Uncovering places to get your questions answered

▶ Locating fun math things to do with your child

A s a parent who is interested in your child's math learning, you may not know how to take action (beyond buying and reading this book!). For literacy, parents are told a simple, clear message: "Read 20 minutes every day with your child." Unfortunately the message hasn't been as clear for parents about children and math. But the equivalent message is equally simple: "Talk about numbers or shapes every day with your child."

In addition, parents are often concerned about helping their kids with math homework. This chapter provides information about resources that can assist you with both concerns. You can find ideas for engaging your child in thinking and talking about math in fun ways, and you also can find ideas for getting specific help when you and your child need it.

Talking Math with Your Kids

I started the Talking Math with Your Kids website (http://talkingmathwithkids.com) in 2013 to give parents examples of how to support the math learning of their children ages 2 to 10. The ideas apply more broadly than these ages, though, and the website has product reviews, summaries of research, and news items in addition to the examples of everyday math conversations.

Moebius Noodles

This project consists of both a website (www. moebiusnoodles.com) and a book. The project's goal is to build a community of people doing mathematics with young children. The website encourages parents to "imagine turning your child's world into a mathematical playground!" The major audience for Moebius Noodles is homeschooling parents and children, but the adventurous and creative work on the site is accessible and useful for any parent interested in exploring math with their children.

Estimation 180

Andrew Stadel is a middle school teacher in California. He has been helping his students improve their number sense by having them guess each day how many things are in a photograph that he shows them at the beginning of the period. He does this exercise every day of the school year (all 180 days — that's where the name comes from). He has turned this idea and his photographs into a free website, Estimation 180 (www. estimation180.com). The interesting photos have clever themes and connections, and they build in complexity from the early days of the school year to the later ones. You can make these exercises a bedtime activity or an after-school snack conversation.

Visual Patterns

Visual Patterns (www.visualpatterns.org) is another great free resource created for teachers that parents also can use to have fun, low-stakes talks with their children.

Fawn Nguyen teaches middle school and has always approached math in a visual way. She *sees* numbers and patterns, and she helps her students to build this skill for themselves. After all, mathematicians visualize the math they work on, too. She recently began collecting and organizing patterns that she uses in her classroom and those sent to her by friends and colleagues on her website.

Math Educators Stack Exchange

Don't be scared away by the technical-sounding name here. The Math Educators Stack Exchange (http://matheducators. stackexchange.com) is a relatively new online forum for anyone concerned with teaching and learning mathematics. If you're helping your child with math, you're a math educator. Ask a question on the site, and you'll likely get several quick and helpful replies.

Math Forum

Drexel University hosts the Math Forum (http://mathforum. org) and bills it as, "the leading online resource for improving math learning, teaching, and communication since 1992." Perhaps the most famous and useful section of the Math Forum is the Ask Dr. Math section, where you can find answers to age-old questions such as "Why can't I divide by zero?" as well as ask new ones of your own. Ask Dr. Math includes a searchable archive of questions and answers.

YouCubed

This site (http://youcubed.org) is a new project based at Stanford University. This nonprofit gives free and affordable math resources for kindergarten through twelfth grade for educators and parents. The site produces and curates videos, articles, links, and resources for parents, teachers, and students.

Your Public Library

When you start checking children's books out of the library, make sure to throw a shapes book and a numbers book into your bag. Doing so can ensure that you're talking about numbers and shapes from your child's earliest days.

 As your children age, ask your librarian for recommendations of books that have math ideas but that may not scream math in their titles. *A Chicken Stayed for Supper* by Carrie Weston (Holiday House) and *Miss Lina's Ballerinas* by Grace

Maccarone (Feiwel & Friends) are two examples of delightful books that get children thinking mathematically but that aren't found in the math section of the library.

Also, most libraries have homework and study help available a few select hours each week. Don't be ashamed to outsource the homework help. Check your library's website or ask your librarian for information.

Math Munch

Math Munch (http://mathmunch.org) is "a weekly digest of the mathematical Internet." Three math teachers comb the web for interesting tidbits and ideas. You won't find traditional homework help here, but you can find all sorts of amazing and wonderful math things to play with and wonder about. This is a place to go for math-based fun and delight.

Common Core State Standards Initiative

I wrote *Common Core Math For Parents For Dummies* to help you understand the big picture of the standards. Sometimes seeing the details can be helpful. If you long to look under the hood of the Common Core machine, head on over to the official website (www.corestandards.org/math/) where the standards are housed.

The site is easy to navigate; just click on the grade that interests you and then on a topic (for example geometry — the website calls these topics *domains*). Alternatively, you can start by clicking on a domain such as geometry. Then you can see the standards for that domain from all of the grade levels together. This area is a great way to get an answer to a question such as, "In what grade will my child learn to measure angles?"

Chapter 16

Ten (or So) Proven Ways to Support Math at Home

In This Chapter

▶ Getting ideas for doing math together beyond the homework routine

▶ Finding ways to keep your child upbeat about learning math

▶ Doing math together

▶ Understanding your role in setting expectations for success in math

*A*mericans tend to talk about being *math people* and *non-math people*. But nothing could be further from the truth. The messages that you send to your children about their potential for success have a deep impact on the ways your children think about their own abilities and potential for success. This chapter gives you ideas for ways to avoid raising a math-o-phobe and to help your child become successful in math.

Talking about Math Together

Set yourself a goal to ask "how many?," "what shape?," and "how do you know?" at least once a day. Don't worry about asking it at the perfect time. Don't worry about asking a question that's too difficult; your child can take a guess, and then *you* can talk. You don't need to know a lot of math in order to support your child's learning. You may even be a little afraid of math, which is okay. Talking about numbers and shapes with your children is a low-risk activity.

Talk about math together in the car, on the bus, at the dinner table, or while walking the family dog. Here are some simple sample prompts for you to try:

✔ How many crackers (or grapes, or apple slices, or pieces of cheese) would you like for your snack?

✔ How many legs do you think are in your collection of toy animals?

Perusing Your Child's Notebook

Parents tend to rely on two sources of information about what is happening in math class from day-to-day: the child's answer to the question, "What did you do today?" and the homework that is assigned. The first is notoriously unreliable, and the second can be a source of stress.

Another way to get a handle on what is happening in class is to look at your child's notebook and ask questions about what you see there. You can ask when it isn't homework time and with a spirit of inquiry. What you see will bring back memories of learning the same content (perhaps in a different way). Share those memories. Connect with your child for a few minutes. He may start making better and more interesting notes.

Explaining a Solved Problem

Too often, the math that children see is limited to the classroom. You probably solve some kind of math problem (even if it's a very simple one) every day. If you estimate whether you have enough money in your bank account to use your debit card or have enough milk for cereal in the morning, you're solving math problems. You may not notice yourself doing it. If you don't notice it, your child doesn't either.

Make notes to yourself when you think about numbers, quantities, or shapes. Don't worry if your examples seem too simple (or too complicated). Then tell your child about an example when you have a spare moment together. Doing so can help him to see that math is used outside of textbooks.

Playing a Board Game

Board games (and card games) are lovely ways to engage young children's minds. Talk about the game as you play. Make a big deal about what you hope you'll get on the next roll or draw of the card. Talk about how likely it is that you will. Discuss whether you got what you hoped for and the difference between it and what you actually rolled (or drew).

Having Number Talks

A *number talk* is a deliberate conversation where you focus on trying to count things in different ways. It's fun because the goal is to think in several different ways. For example, some grocery stores sell large containers of 18 eggs. You and your child already know that there are 18 (it's written on the box). But the two of you can imagine for a moment that you don't know this (or that you want to check that the store isn't lying to you). How could you count the eggs? You could count them one at a time. You could see three rows of six or six columns of three. There are many possibilities.

You can have a number talk to count many images and situations. The more you look for them, the more you'll see. The more numbers talks you have, the better you'll get at seeing different structures and groupings — and your child will too.

Grocery Shopping Together

The grocery store has a wealth of opportunities to discuss numbers and shapes. Here are some examples:

- ✔ Do all the jam, jelly, and mustard jars have circular bottoms? If not, what other shapes can we find?

- ✔ If you're paying cash, which bills and coins should you pay with? How much change should you get back?

- ✔ Is a $3, 24-ounce jar of peanut butter a better or worse deal than a $2, 16-ounce jar? How do you know?

Arguing about Words

Mathematicians use words much more precisely in their work than most people do in their everyday lives. When your 8-year-old wants to argue that he isn't technically *jumping* on the couch because he is on his knees, keeping your cool can be difficult. But if you manage to do so, you can engage him in an important intellectual activity by asking "What would you call what you're doing?" and by building your own argument, "I would call it jumping anytime you leave the surface you are resting on." One or two rounds of back-and-forth before you tell him to play outside until dinnertime will sharpen his mind for the definitions he'll have to learn and to write for himself in math class. *Polygon*, *number*, and *fraction* are all examples of math words that can be tricky to define precisely.

Building Things

Spatial visualization is an important skill for developing an understanding of geometry and (perhaps surprisingly) other areas of math as well. One of the best ways for your child to develop spatial visualization skills is to build things. You can provide blocks, boxes, sugar cubes, toilet paper rolls, cardboard, whatever. When your child imagines how he wants the structure to look and then compares his imagination to the reality, he will develop his spatial visualization skills. You can build together with your child, or he can build alone or with friends. What matters is that he gets time to imagine spatial relationships as he plays.

Asking "What If?"

Important math discoveries come from asking "What if?" If you make a regular habit of asking these types of questions together with your child, especially having to do with numbers and shapes, he will also make a habit of thinking beyond the possibilities that he encounters from teachers and in his textbook. He may even go on to make new mathematical discoveries.

Index

• A •

accountability, of teachers, 21–23
ACT score, 169
addition
    in first-grade math, 57–62
    fractions, 115–116
    in order of operations, 109
    in second-grade math, 70–72
    in third-grade math, 85–86
additive structure, 30
adjacent angles, 148, 149
advanced math
    about, 167
    algebra in, 172–176
    calculus, 171–172
    college and career ready, 167–171
    college math placement, 168–169
    geometry, 177–180
    integrated math, 170–171
    statistics, 180–184
algebra
    in advanced math, 172–176
    in fifth-grade math, 106–111
    in kindergarten math, 50–51
    in seventh-grade math, 145–146
    in sixth-grade math, 130–131
Algebra 2 For Dummies
    (Sterling), 173
algebraic structure, in
    advanced math, 176
analytic geometry, 177
angles
    in fourth-grade math, 102–104
    types of, 148, 149
area
    counting squares to find, 91–92
    measuring, 132–134
area models, 98

arguing
    about, 10–11
    about words, 194
    in math, 34–35
arrays, 73–74
asking
    "Does this make sense?," 29–30
    "How do you know?," 26–27
    "Is it good enough?," 28–29
    questions, 10
    "What if?," 27, 194
    "What's going on here?," 30–32
    "Why?," 26–27
associative property, 83, 144
attending to precision, 28

• B •

base-ten blocks, 63
base-ten-place value system, in
    first-grade math, 62–65
benchmarks, 77, 100–101
bimodal, 136
bivariate data, in eighth-grade
    math, 166
board games, playing, 193
borrowing, 63, 75
building
    models, 33–34
    shapes, 78

• C •

calculating, with place value, 94–96
calculus, 171–172
cardinality, 47–49
carrying, 63, 75
categorizing shapes, 92

center, measures of, 135–136
chance, in seventh-grade math,
    149–150
Cheat Sheet (website), 4
A Chicken Stayed for Supper
    (Weston), 189
circles
  in first-grade math, 64
  measuring, 147
  in seventh-grade math, 147
circumference, 147
classifying, in kindergarten math,
    52–53
college and career ready, 167–171
college math placement, 168–169
college-level math, 169
Committee of Ten, 17–18
Common Core Math. *See also*
    specific topics
  about, 7–8
  helping with homework, 16
  Standards by grade, 11–15
  Standards for Mathematical
    Practice, 9–11, 25
  what it is, 8–9
Common Core Standards, release
    of, 23–24
Common Core State Standards
    (CCSS), 23
Common Core State Standards
    Initiative, 190
common denominator algorithm,
    128
common factors, 125–126
common multiples, 125–126
commutative property, 73,
    82–83
comparing
  denominators, 99–100
  fractions, 99–101
  in kindergarten math, 52–53
  numbers, 49–50
  numerators, 100
complementary angles, 148
composing, 56

composite number, 94
computational fluency, 95–96
congruence
  about, 148, 149
  compared with similar, 162
  in eighth-grade math, 162–163
connecting, 11, 35–36
cosecant, 179
cosine, 179
cotangent, 179
Council of Chief State School
    Officers (CCSSO), 23
counting
  about, 47–49
  by ones and tens, 47–50
  solving problems by, 50–51
  squares to find area, 91–92
cross-multiply, 141–142
cubes, linking, 63
*Curriculum and Evaluation Standards
    for School Mathematics (NCTM
    Standards)*, 20–21

• D •

data
  bivariate, 166
  graphing, 118
  measuring, 67
  quantifying, 181–182
  relationship with probability,
    182–183
  representing with graphs, 92
  in seventh-grade math,
    149–150
data analysis, 181
datasets, measuring, 135–136
decimal number system, in
    first-grade math, 62–65
decimal place value, 111–112
decision-making, with
    probability, 183–184
decomposing
  about, 56
  numbers, 61–62, 95

defining attributes, 67
denominators, 96, 99–100
describing shapes, 54–55, 179–180
diameter, 147
Dienes, Zoltan (mathematician), 63
Dienes blocks, 63
difference, 138, 143
digits, 74
distance, 142
distributive property, 83–84,
   131, 144
division
   fractions, 126–130
   negative numbers with
      properties, 143–144
   in order of operations, 109
   in third-grade math, 79–84
"Does this make sense?," asking,
   29–30
domains, 12, 190
dots, 41–42
Dummies (website), 4

• E •

eighth-grade math
   about, 151
   bivariate data, 166
   congruence, 162–163
   exponentiation, 156–157
   exponents equations, 152–159
   functions, 159–162
   geometry, 162–166
   irrational numbers, 151–152
   linear relationships, 153–155
   measuring hypotenuse, 163–165
   measuring volume,
      165–166
   Standards for, 15
   system of equations, 157–159
elimination, 175
'equal,' 49–50
equivalence, in sixth-grade
   math, 131

equivalent fractions
   about, 124
   in fourth-grade math, 97–99
estimating, in third-grade
   math, 90–92
Estimation 180, 188
even number, 106
expected value, 183
explaining solved
   problems, 192
exponentiation
   in eighth-grade math, 156–157
   in order of operations, 109
exponents equations, 152–159
expressions, 106–108

• F •

factorial, 110
factors, 80, 93–94
fifth-grade math
   about, 105
   algebra, 106–111
   fractions, 114–117
   graphing data, 118
   measuring volume, 117–118
   order of operations, 108–111
   place value, 111–112
   properties of shapes, 118
   standard algorithms, 112–114
   Standards for, 14
first-grade math
   about, 57
   addition, 57–62
   circles, 64
   data, 67
   decomposing numbers, 61–62
   geometry, 67–68
   keyword strategy, 58–60
   length, 65–66
   measurements, 65–67
   money, 66
   place value, 62–65
   representing tens, 63–65

first-grade math *(continued)*
significance of 10, 62–63
Standards for, 12–13
subtraction, 57–62
time, 66
fluency, 95
formative assessment, 43–44
fourth-grade math
about, 93
angles, 102–104
calculating with place value,
94–96
comparing fractions, 99–101
computational fluency, 95–96
decomposing numbers, 95
equivalent fractions, 97–99
factors, 93–94
fractions, 96–101
lines, 103–104
multiples, 93–94
multiplication, 93–94
Standards for, 13–14
unit fractions, 96–97
units of measure, 101–102
fractions
adding, 115–116
compared with ratios, 122–123
comparing, 99–101
dividing, 126–130
equivalent, 97–99, 124
in fifth-grade math, 114–117
in fourth-grade math, 96–101
multiplying, 116
reducing, 97–98
subtracting, 115–116
in third-grade math, 87–90, 89–90
function families, 173–174
functions
in eighth-grade math, 159–162
functions of, 161–162

• G •

generalized arithmetic, 107
generalizing, 107

geometry
in advanced math, 177–180
in eighth-grade math, 162–166
in first-grade math, 67–68
in kindergarten math, 53–56
in seventh-grade math, 146–149
grade, Standards by, 11–15
graphing data, 118
graphs, representing data
with, 92
greatest common factor (GCF),
125–126
grocery shopping, 193
groups, in second-grade math,
72–74

• H •

helping, with homework, 16
hierarchy of quadrilaterals, 118
high school, Standards for, 15
high school GPA, 169
high school math. *See*
advanced math
homework
about, 37
dots, 41–42
formative assessment, 43–44
helping with, 16
number lines, 39–41
strategy for, 42–43
teacher views on, 37–38
"How do you know?,"
asking, 26–27
hundreds chart, in first-grade
math, 64–65
hyperbola, 180
hypotenuse, measuring, 163–165

• I •

icons, explained, 3–4
ideas, connecting, 35–36
identifying shapes, 78
implied multiplication, 109

information age, math
    teaching in the, 20–21
integrated math, 170–171
invert, 128–130
irrational numbers, 151–152
"Is it good enough?," asking, 28–29
iterating, 66, 76

• K •

keyword strategy, in first-grade
    math, 58–60
kindergarten math
    about, 47
    algebra, 50–51
    cardinality, 47–49
    classifying, 52–53
    comparing, 52–53
    comparing numbers, 49–50
    counting, 47–49
    counting by ones and tens,
        47–50
    describing shapes, 54–55
    geometry, 53–56
    place value, 52
    playing with shapes, 55–56
    Standards for, 12

• L •

least common multiple (LCM),
    125–126
length, measuring, 65–66, 76–77
'less,' 49–50
libraries, public, 189–190
line of symmetry, 104
line plot, 118
line segments, 103–104
linear relationships
    about, 15
    in eighth-grade math, 153–155
lines
    in first-grade math, 64
    in fourth-grade math, 103–104

linking cubes, 63
long division, 113

• M •

Maccarone, Grace (author)
    Miss Lina's Ballerinas, 189–190
matching correspondence, 173
math. *See also specific types*
    arguments in, 34–35
    home tips for supporting,
        191–194
    talking about, 191–192
Math Educators Stack Exchange,
    189
Math Forum, 189
Math Munch, 190
math teaching
    1970s, 19–20
    accountability of teachers, 21–23
    Common Core Standards
        released, 23–24
    competing globally, 18–19
    goals in 1900s, 17–18
    in the information age, 20–21
    playing with, 32–34
Math Wars, 21
mathematical modeling, 176
Mathematical Practice, Standards
    for, 9–11
mean, 135, 181
mean absolute deviation (MAD),
    136
measures of center, 135–136
measuring
    area, 132–134
    circles, 147
    data, 67
    datasets, 135–136
    in first-grade math, 65–67
    hypotenuse, 163–165
    in second grade math, 76–77
    in third-grade math, 90–92
    volume, 117–118, 134, 165–166

measuring division, 126
measuring problem, 80–81
median, 135, 181
mental math, 51
Miss Lina's Ballerinas (Maccarone), 189–190
mode, 135–136, 181
models, building, 33–34
Moebius Noodles, 188
money
  in first-grade math, 66
  in second-grade math, 77
'more,' 49–50
multiples, in fourth-grade math, 93–94
multiplication
  commutative property of, 73
  in fourth-grade math, 93–94
  fractions, 116
  negative numbers with properties, 143–144
  in order of operations, 109
  in sixth-grade math, 125–130
  in third-grade math, 79–84
multiplicative structure, 30

● *N* ●

National Council of Teachers of Mathematics (NCTM), 20
National Education Association (NEA), 18
National Governors Association (NGA), 23
negative numbers, in seventh-grade math, 142–144
New Math, 19
Nguyen, Fawn (teacher), 188
No Child Left Behind (NCLB), 11, 22–23
number facts, knowing, 70
number lines
  about, 39–41
  negative numbers and, 142–143
  in third-grade math, 89–90

number talks, 193
numbers
  comparing, 49–50
  composite, 94
  decomposing, 61–62, 95
  even, 106
  irrational, 151–152
  odd, 106
numerators, 87, 96, 100

● *O* ●

odd number, 106
ones, counting by, 47–50
online material, 4
operations, 12
order of operations, in fifth-grade math, 108–111

● *P* ●

pair, 72
parabola, 171
parallel lines, 104
parentheses, in order of operations, 109
parents, resources for, 187–190
partitioning
  about, 78
  in third-grade math, 87–89
PEMDAS mnemonic, 109–110
perimeter, 92
perpendicular lines, 104
pi, in seventh-grade math, 147
place value
  about, 13
  calculating with, 94–96
  defined, 69
  in fifth-grade math, 111–112
  in first-grade math, 62–65
  in fourth-grade math, 94–96
  in kindergarten math, 52
  in second grade math, 74–76
place value number system, 74
placement tests, 169

place-value blocks, 63
playing
  about, 10
  board games, 193
  with math, 32–34
  with shapes, 55–56
point, 142
prime number, 94
probability
  decision-making with,
    183–184
  defined, 149
  relationship with data, 182–183
  in seventh-grade math,
    149–150
product, 80
proofs, 177–178
proper factors, 94
properties
  associative, 83, 144
  commutative, 73, 82–83
  distributive, 82–83, 131, 144
  of shapes, 118
  in third-grade math, 82–84
proportional, 178
proportional relationships
  about, 15
  in seventh-grade math, 138–142
proportions
  defined, 139
  solving, 141–142
proving theorems, 178
public libraries, 189–190
'putting together,' 50–51
Pythagorean theorem, 163–165

• *Q* •

quadratic structure, 176
quadrilateral, 92, 118
quantifying data, 181–182
quarters, 87
questions
  asking, 10
  focus on, 26–32

• *R* •

Race to the Top program, 24
radius, 147
range, 136
rate, 139
rate problems, solving, 123–124
ratio table, 124
ratios
  about, 122
  compared with fractions, 122–123
  defined, 138–139
  in seventh-grade math, 138–142
  solving problems, 123–124
ray, 102
reciprocal of a fraction, 128–129
rectangular prism, 117
reducing fractions, 97–98
regrouping, 63
relationships between variables, 173
Remember icon, 3
representing
  data with graphs, 92
  tens, 63–65
resources, for parents, 187–190

• *S* •

SAT score, 169
scatter plot, 182
secant, 179
second-grade math
  about, 69
  addition, 70–72
  groups, 72–74
  measurement, 76–77
  place value, 74–76
  shapes, 78
  Standards for, 13
  subtraction, 70–72
seventh-grade math
  about, 137
  algebra, 145–146
  chance, 149–150
  circles, 147

seventh-grade math *(continued)*
data, 149–150
geometry, 146–149
negative numbers, 142–144
pi, 147
probability, 149–150
proportional relationships,
138–142
ratios, 138–142
Standards for, 15
statistics, 150
triangles, 147–149
shapes
building, 78
categorizing, 92
describing, 54–55, 179–180
identifying, 78
playing with, 55–56
properties of, 118
in second grade math, 78
sharing division, 126
sharing problem, 80–81
similarity
in advanced math, 178–179
compared with congruence, 162
simplifying, 98
sine, 179
sixth-grade math
about, 121
algebra, 130–131
comparing fractions and ratios,
122–123
comparing GCFs and LCMs,
125–126
dividing fractions, 126–130
equivalence, 131
measures of center, 135–136
measuring area, 132–134
measuring datasets, 135–136
measuring volume, 134
multiplication, 125–130
ratios, 122–124
solving ratio and rate problems,
123–124
spread, 136

Standards for, 14
variables, 130–131
skip counting, 66, 80, 88
solved problems, explaining, 192
solving
problems by counting, 50–51
proportions, 141–142
rate problems, 123–124
ratio problems, 123–124
systems, 174–175
spatial visualization
about, 55–56, 194
in first-grade math, 68
spread, 136
squares, counting to find area,
91–92
Stadel, Andrew (teacher), 188
standard algorithms
about, 96
in fifth-grade math, 112–114
standard deviation, 136, 181
Standards for Mathematical
Practice (SMP)
about, 9–11, 25
focus on asking questions,
26–32
by grade, 11–15
statistical inference, 150
statistics
in advanced math, 180–184
in seventh-grade math, 150
Sterling, Mary Jane (author)
Algebra 2 For Dummies, 173
story problems, 145–146
strategy, for helping with
homework, 42–43
structure, 30, 176
structure of computations, 106
substitution, 175
subtraction
in first-grade math, 57–62
fractions, 115–116
in order of operations, 109
in second-grade math, 70–72
in third-grade math, 85–86

supplementary angles, 148, 149
symbols, 32–33
symmetry, line of, 104
system of equations
about, 174
in eighth-grade math, 157–159
system of inequalities, 174
systems, solving, 174–175

• T •

'taking apart,' 50–51
talking about math, 191–192
Talking Math with Your Kids, 187–188
tangent, 179
teachers
accountability of, 21–23
views on homework of, 37–38
teaching. *See* math teaching
Technical Stuff icon, 4
ten frame
about, 52
significance of, 62–63
tens
counting by, 47–50
in kindergarten math, 52
representing, 63–65
third-grade math
about, 79
addition, 85–86
categorizing shapes, 92
counting squares to find area, 91–92
division, 79–84
estimating, 90–92
fractions, 87–90
measuring, 90–92
multiplication, 79–84
number lines and fractions, 89–90
partitioning, 87–89
properties, 82–84
representing data with graphs, 92

Standards for, 13
subtraction, 85–86
tilty triangle, 134
time
in first-grade math, 66
in second grade math, 77
Tip icon, 3
trading, 63
transformational geometry, 177
triangles, in seventh-grade math, 147–149
trigonometry, in advanced math, 178–179
Try This icon, 3

• U •

unit, 72, 87
unit fractions, 13, 87–89, 96–97, 115–116
unit rate, 139, 141
units of measure, in fourth-grade math, 101–102

• V •

variables
quantifying data in one or two, 181–182
relationships between, 173
in sixth-grade math, 130–131
vertex, 102
vertical angles, 148, 149
Visual Patterns, 188
volume, measuring, 117–118, 134, 165–166

• W •

websites
Cheat Sheet, 4
Common Core State Standards Initiative, 190
Dummies, 4

websites *(continued)*
Estimation 180, 188
Math Educators Stack
Exchange, 189
Math Forum, 189
Math Munch, 190
Moebius Noodles, 188
online material, 4
Talking Math with Your Kids, 187
Visual Patterns, 188
YouCubed, 189

Weston, Carrie (author)
A Chicken Stayed for Supper, 189
"What if?," asking, 27, 194
"What's going on here?," asking,
30–32
"Why?," asking, 26–27
word problems, 145–146

YouCubed, 189

About the Author

Christopher Danielson began his career teaching seventh-grade math in the St. Paul Public Schools. He went on to earn his PhD in mathematics education from the Department of Mathematics at Michigan State University and now teaches at Normandale Community College in Bloomington, Minnesota. He teaches college algebra and calculus as well as math content courses for future elementary and special education teachers. His blog, Talking Math with Your Kids, is dedicated to helping parents support their children's mathematical growth.

Dedication

I dedicate this book to my dear wife Rachel. She didn't bat an eye when I told her that I would be taking on this project, and she has been supportive of all of my professional and learning opportunities that have led to being ready to write this book.

Author's Acknowledgments

I want to thank my editors for their hard work and support. Lindsay Lefevere, my acquisitions editor, was enthusiastic and supportive from the beginning. Chad Sievers, my project editor, has been a tireless answerer of my many questions.

I also want to thank the many communities of teachers who have supported my work over the years, including those who convene in Michigan each summer, my colleagues teaching K–12 and beyond throughout the state of Minnesota, my online colleagues who make me think, and my colleagues and administration at Normandale Community College. Thank you all for making me a better teacher.

I want to thank Griffin and Tabitha, and I want to thank Jennifer Madden and Anne Bartel.

Publisher's Acknowledgments

Executive Editor: Lindsay Lefevere

Project Editor: Chad R. Sievers

Copy Editor: Chad R. Sievers

Technical Editors: Leslie Sanford Arceneaux and Pam Goodner

Art Coordinator: Alicia B. South

Illustrations: Thomson Digital

Project Coordinator: Sheree Montgomery

Cover Photos: ©iStock.com/eli_asenova